Rotation: The Key to Understanding Stars

Sana

Contents

1 Introduction

Introduction   1

1.1 Stellar rotation period . . . .
    1.1.1 Rotation evolution – Age connection . . . .
    1.1.2 Skumanich law and deviations . . . .
    1.1.3 Stellar Activity & Rotation Period . . . .
1.2 Solar and stellar brightness variability . . . .
1.3 Initial solar and stellar photometric records. . . . .
1.4 Space born photometry . . . .
    1.4.1 Total Solar Irradiance . . . .
    1.4.2 Stellar photometric data . . . .
        1.4.2.1 *Kepler* mission . . . .
    1.4.3 Rotation period in planetary transits analysis . . . .
1.5 Photometric methods for rotation periods detection . . . .
    1.5.1 Generalized Lomb-Scargle periodogram – (GLS) . . . .
    1.5.2 Auto-Correlation Functions – (ACF) . . . .
    1.5.3 Wavelet Power Spectra – (PS) . . . .
    1.5.4 Gaussian Process – (GP) . . . .
    1.5.5 Gradient of the Power Spectra: GPS . . . .
1.6 State of the Art . . . .
1.7 The Sun is less active than other solar-like stars . . . .
    1.7.1 Rvar distribution . . . .
1.8 Decoding the radial velocity variations of HD41248 with ESPRESSO . .
    1.8.1 TESS . . . .
    1.8.2 Analysis of the TESS light curve . . . .
    1.8.3 Rotation period . . . .
1.9 The correlation between photometric variability and radial velocity jitter .
    1.9.1 Light curve: TESS . . . .
    1.9.2 Stellar rotation period . . . .
    1.9.3 Correlation with stellar rotation period . . . .

2 Inflection point in the power spectrum of stellar brightness variations: I. The model 32

2.1 Introduction of chapter 2 . . . .

2.2 Model description . . . .

2.3 Stars with spots . . . .

    2.3.1 High-frequency tail of the power spectrum and inflection point . .

    2.3.2 Effect of spot emergence and lifetime . . . .

2.4 Stars with spots and faculae . . . .

    2.4.1 Treatment of faculae . . . .

    2.4.2 Superposition of spot and facular contributions to stellar brightness variability . . . .

2.5 Main factors affecting position of the inflection point . . . .

    2.5.1 Position of the inflection point as a function of the facular to spot area ratio . . . .

    2.5.2 Inflection point in the power spectrum of solar brightness variations

    2.5.3 Effect of inclination . . . .

2.6 Position of the inflection point as a function of stellar magnetic activity . .

2.7 Conclusions . . . .

2.8 Appendix chapter 2 . . . .

    2.8.1 Additional figures . . . .

    2.8.2 Solar values of facular to spot area ratio at the time of maximum area

    2.8.3 Examples of testing GPS method on *Kepler* stars . . . .

3 Inflection point in the power spectrum of stellar brightness variations. II. The Sun 69

3.1 Introduction of chapter 3 . . . .

3.2 Methods for determining stellar-rotation periods . . . .

3.3 Validation of GPS method for the solar case . . . .

    3.3.1 Records of total solar irradiance, TSI . . . .

    3.3.2 Brightness signature of spot transits . . . .

    3.3.3 Brightness signature of facular feature transits . . . .

    3.3.4 Analysis of the entire data-set . . . .

    3.3.5 The solar variability in 90-day quarters . . . .

    3.3.6 The impact of white noise in the inflection point position . . . . . .

3.4 GPS and skewness relation . . . .

3.5 Discussion & Summary . . . .

4 Inflection point in the power spectrum of stellar brightness variations III: Facular versus spot dominance on stars with known rotation periods 92

4.1 Introduction of chapter 4 . . . .

4.2 Stellar sample selection . . . .

4.3 Results and discussion . . . .

4.4 Summary . . . .

4.5 Appendix chapter 4 . . . .

Summary & outlook 110

1 Introduction

1.1 Stellar rotation period

Understanding the physics behind magnetic activity in stars is a challenging task, even when such stars are analogs to the most studied star, our Sun. There are many variables and degeneracies working simultaneously in the attempt to recover magnetic properties of stars. But, there is a key parameter for characterising the physics behind magnetic activity, this parameter is the rotational period. Accurate surveys of rotational periods are crucial for the understanding of stellar dynamo theory. As well, important for a better tracing of stellar evolution and age calibration (Ulrich 1986; Barnes 2003).

The importance of rotation period information goes beyond the understanding of fundamental properties. That information bring us close to the understanding of phenomena such as, stellar structure, mixing and interior processes, light elements evolution (Be, Li), the star accretion/formation and disk/planets interaction, angular and mass loss rates, stellar ages, history/future of activity, magnetic field generation, etc. All in all, precise rotation periods are needed for recovering stellar information to properly characterise stars.

1.1.1 Rotation evolution – Age connection

Stars are born after the gravitational collapse of a molecular cloud leading to a hot ball of plasma with an initial angular momentum. This is then when the stellar rotation evolution begins. Multiple processes present in the stars are drastically affected by the stellar rotation evolution, as activity and the dynamic magnetism (see Bouvier 2013; Gallet and Bouvier 2013; Gallet et al. 2019).

The rotation evolution strongly depends on the stellar initial conditions. During the proto-stellar stage (0.1 - 5 Myr, depending on the initial mass) the amount of dust in the proto-stellar cloud, the accretion rate and, the initial angular momentum play an important role in the evolution path that the rotation of the star will follow.

During the pre-main sequence (PMS, about 5 - 40 Myr) stars gain mass accreting from the remaining molecular cloud. Stars will spin up and increase their angular momentum. After this process the accretion slows-down and a disk of dust is generated around the star. Then, a strong interaction between the stellar magnetic field and the disk of material will diminish the stellar angular momentum. The star will spin-down, and this will mark the the beginning of the main-sequence stage (40 - 10.000 Myr, for Sun-like stars, see Fig. 1.1).

The reconstruction of the rotation period evolution give us important keys and direct information for the tracing of stellar ages.

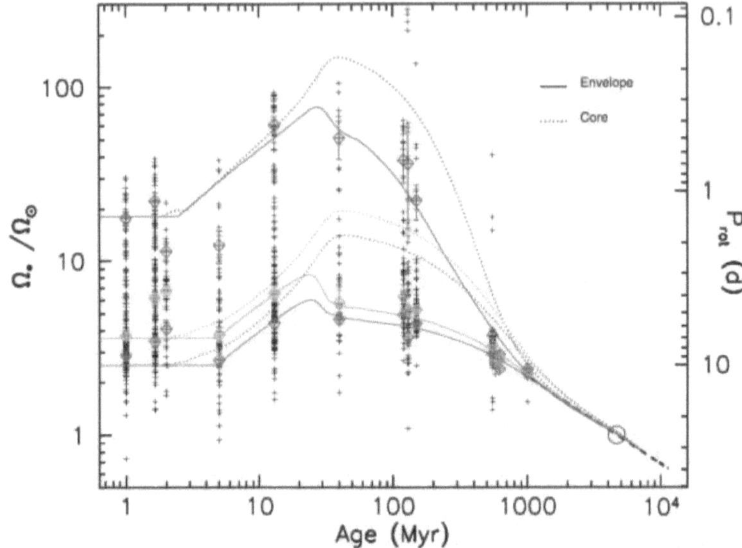

Figure 1.1: Rotation/angular momentum evolution model for Sun-like stars. The figure illustrates different angular velocities distributions of solar-like stars observed in open clusters from 1 to 1000 Myr. The model shows the rotation of the radiative core and convective envelope (dash and solid lines respectively). The Skumanish law is drawn with a dashed black line. The Sun is represented with an open circle (see Gallet and Bouvier 2013).

1.1.2   Skumanich law and deviations

In 1972 Skumanich measured rotation periods, the amount of emission in Ca II H and K lines and, the lithium abundances in a number of stars from the Pleiades, Hyades and, Ursa major clusters. He found that the amount of emission in Ca II H and K lines drops with the age. Furthermore, he found that lithium abundance follows the same trend as Ca II H and K emission, see Fig.1.2. He concluded that the angular velocity of a star is inversely proportional to the square root of stellar age, i.e., $\Omega \propto t^{-1/2}$. Since then this is known as the Skumanich law, which builds a foundation of gyrochronology, i.e. determination of stellar age from rotation period (see, e.g. Soderblom and Mayor 1993; Krishnamurthi et al. 1997; Barnes 2007).

Even though Skumanich law works quite well for many stars, some other targets from the analysed stellar clusters appear to rotate faster than expected from their age, see Fig.1.3. The observed bi-modality of fast and slow rotators in the same cluster suggests that there may be two possible mechanisms for the stars to spin down. In Sun-like stars the principal mechanisms of angular momentum loss and spin down is the interaction between the stellar magnetic field and the ionised material carried in the stellar wind. A second possible mechanism is showed in Garraffo et al. (2018), where they proposed that surface magnetic field morphology has a strong influence on wind-driven angular momentum loss and, the bi-modal distribution of rotation periods observed in young open clusters (OCs) could be explained by their models. Their predictive model show how a different magnetic complexity configuration can account for a different stellar spin-down path. They attribute the rotation period bifurcation to different stellar magnetic field configurations and its relative interactions with the stellar wind.

There are still many open questions on the way to understand the relation between stellar age and rotation period. Some of them might be caused by inability to properly measure stellar rotation periods. Therefore, accurate surveys of rotation periods can open new windows to clarifying the current picture and solving problems.

1.1.3 Stellar Activity & Rotation Period

Across the Hertzsprung-Russell diagram (henceforth HRD) stars are known to manifest their activity through different observable phenomena over the entire electromagnetic spectrum. Evidence of magnetic activity can be traced from X-rays to radio waves. From fast stellar winds in hot stars generating strong shock-heating R-rays emissions to radio outburst

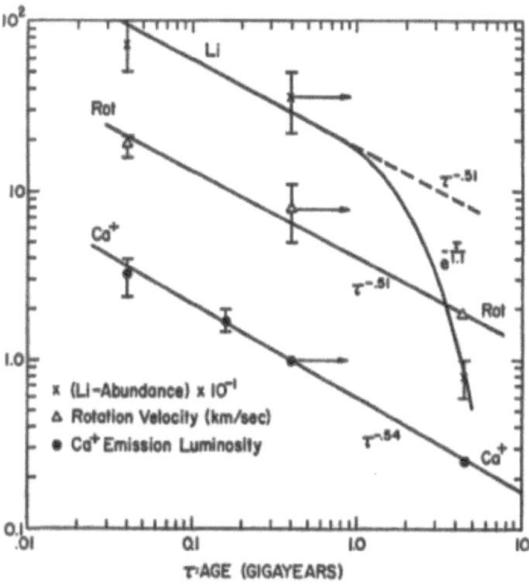

Figure 1.2: Calcium emission, rotation period, and lithium abundance versus stellar age (taken from Skumanich 1972).

generated by flares and CME's in the Sun (see Gopalswamy et al. 2005) and cool dwarfs (F- to M-Type stars). Stellar behaviour has been studied and described in detail over the entire HRD (see, e.g., right panel of Fig. 1.4 for stellar characterisation by radio luminosities). For the scope of the book we will concentrate on activity behaviour of stars in the cool tail of the HRD, more specifically on F- to K- late type main sequence stars.

Concentrations of strong localised magnetic fields emerge in the stellar surface leading to the formation of photospheric magnetic features, such as bright faculae and dark spots (Solanki et al. 2006). The transits of these co-rotating inhomogeneities over the visible disk as the star rotates imprints particular patterns into the observed light-curve. Those characteristics are well associated to stellar activity and provide a way for tracing stellar rotational period.

Rotation period information is essential for determining the action of stellar dynamo, transport of the magnetic flux through the convective zone, and its emergence over the stellar surface, (see for a general review of dynamo theory Charbonneau 2010). The

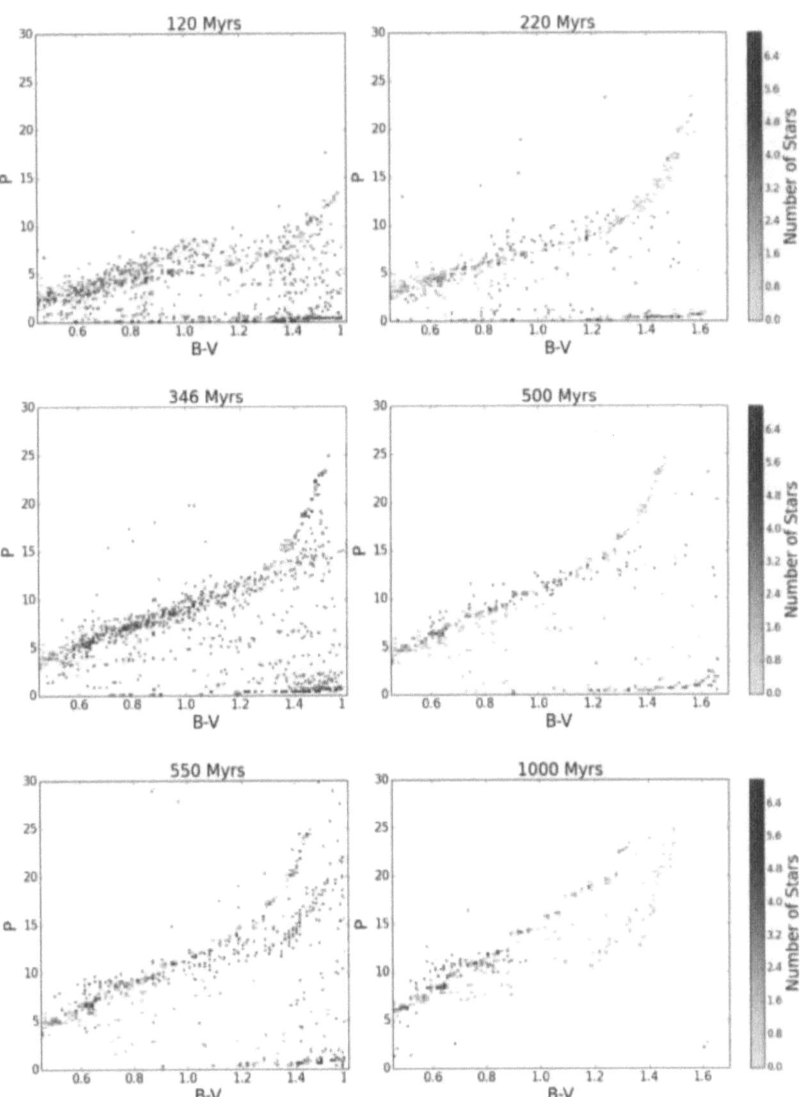

Figure 1.3: Stellar rotation period versus the (B-V) magnitude observed in different open clusters from 120 to 1000 Myr are shown in red. Modelled density distribution of predicted rotation evolution by (see Garraffo et al. 2018) are shown in blue.

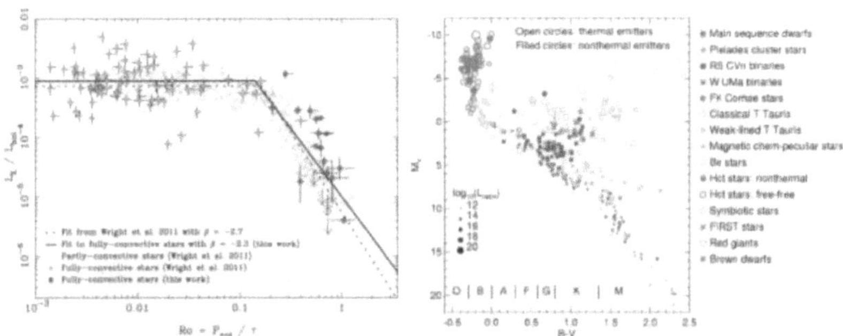

Figure 1.4: Left: Stellar rotation-activity relationship for partially and fully convective stars. X-ray to bolometric luminosity ratio plotted against the Rossby number. Fully convective stars (large red points), and stars in the sample of Wright et al. (2011) (medium, light red points). Remaining partly convective stars from that sample (grey empty circles), find entire description in Wright et al. (2018). Right: Radio luminosities in the Hertzsprung-Russell diagram, taken from Maria Massi lectures.[1]

stellar activity in convective stars can be traced in X-rays, that accounts for the coronal behaviour. In slow rotating stars, with a Rossby number about 0.5 and higher – also called on unsaturated regime – the X-rays luminosities are correlated with the rotation period of the star. On unsaturated regime the magnetic activity of the chromosphere (as indicated by the Ca II K line-core emission) and the corona (as indicated by the X-ray emission) monotonically increase with the stellar rotation rate (see Pizzolato et al. 2003; Wright et al. 2011; Reiners 2012; Reiners et al. 2014). Still, X-rays and high energy observations are non-trivial and are very scarce on stars different than the Sun. Consequently, the rotation period appears to be not just a good raconteur of stellar activity but easier to retrieve tracer of magnetic activity, see the activity rotation diagram at left diagram in Fig. 1.4. Surprisingly, the relationship of rotation period with coronal and chromospheric activity works also for slowly-rotating fully convective stars, (see Wright and Drake 2016; Newton et al. 2017; Wright et al. 2018).

While the modulation of the brightness amplitude is periodic in most of the stars with high and moderate activity levels, patterns on light-curves from magnetic structures of slow rotators as the Sun are quasi-periodic and irregular. Low variability amplitude, short life-time magnetic features evolution (in comparison to the rotation time-scale), that generates irregular modulation on the light-curves, are the main cause of unreliable determination of rotational periodicity in the Sun and its closer analogs. The indistinguibility of magnetic features latitudinal location, clumping or nesting of features, differential rotation and stellar inclination, are additional degeneracies to concern when we want to describe in a simple model the physics under the non-periodic brightness variability (see Isik et al. 2018). The work in this book is focused on the analysis of stellar brightness variations for recovering rotation periods.

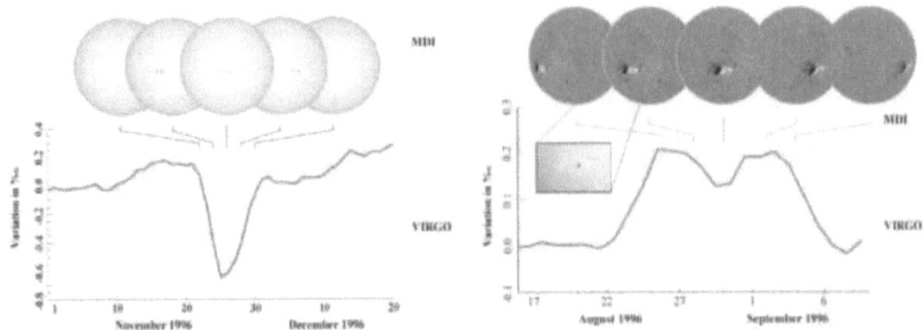

Figure 1.5: TOP: Intensitigram of a spot transiting the solar disk (Left) and magnetogram showing the a faculae transiting (right) by the Michelson Doppler Imager (MDI) instrument. Bottom: Simultaneous to up-images VIRGO TSI variations. Taken from Fligge et al. (2000a)

1.2   Solar and stellar brightness variability

The Sun is by far the star with the most recorded and analysed data. Direct images, spectroscopy, polarimetry, photometry in different bands of the electromagnetic spectrum are available to describe its behaviour. The understanding of physical processes on the Sun is the benchmark and main guide in order to characterise similar phenomena in other analog stars. Performing a one to one comparison between the Sun and other stars is, however, challenging. The differences are mostly set by observational constrains; the Sun observed from Earth is a resolved object, which is mostly not the case for other Sun-like stars. Also the cadence of the solar and stellar measurements is pretty different. For a long time solar data have been degraded to the cadence of stellar observations (see, e.g. Lockwood et al. (1997)). On the contrary, now we are in a unique situation that stellar photometric data have often better cadence than the solar ones.

Solar brightness variability has been associated with stellar surface processes on different time-scales, from granulation to the formation of magnetic features. Those processes, help us to determine the activity cycles, passing through rotational modulation.

The variability of solar brightness is one of the most intriguing manifestations of its magnetic activity. Magnetic field emerges on the solar surface in the form of flux concentrations and leads to the formation of bright facular and dark spot magnetic features. In Chapman et al. (1997) determined that bright structures contributes to the irradiance excess, associated for example with facular regions, by outweighs about 50% the irradiance deficit associated with sunspots. These features imprint very different pattern into the solar light curve (see, Fig. 1.5).

Imprints from spot and/or facular components on light-curves bring a handle tool to describe stellar surface and to interpret rotation (see, Reinhold and Reiners 2013; Shapiro et al. 2016, 2017). The latter is a key parameter to define the stellar dynamo mechanism, the transport of magnetic flux through the convective zone, and its emergence on the stellar surface, among other phenomena (see, Charbonneau 2010; Reiners et al. 2014; Fabbian et al. 2017).

1.3   Initial solar and stellar photometric records.

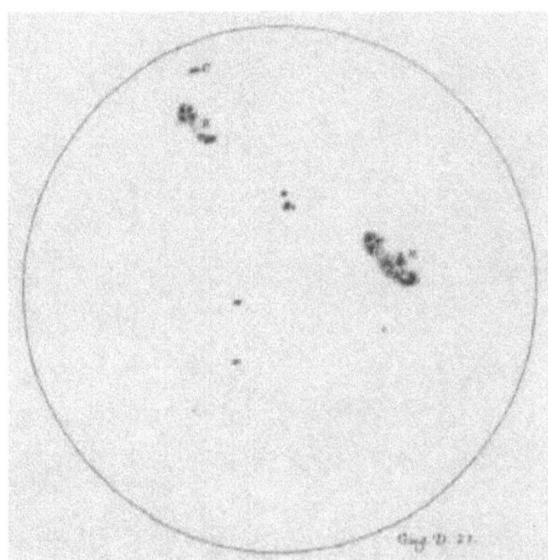

Figure 1.6:   Drawings of sunspots by Galileo Galilei reported by Giacomo Mascardi in History and Demonstrations Concerning Sunspots and their Properties Mascardi (1613): *"Istoria e dimostrazioni intorno alle macchie solari e loro accidenti comprese in tre lettere scritte all'illustrissimo signor Marco Velseri linceo ... dal signor Galileo Galilei linceo ... Si aggiungono nel fine le lettere, e disquisizioni del finto Apelle"*: *"I am at last convinced that the spots are objects close to the surface of the solar globe ... also that they are carried the Sun by its rotation ..."* **–Galileo.**

The rotation period of the Sun could by traced by observing the transit of magnetic features. Already in 1612 following of solar surface performed the first observations and drawings of sunspots (published in Mascardi 1613, see Fig. 1.6). In 1863 Richard Christopher Carrington published his book: *Observations of the Spots on the Sun*, (Carrington 1863), where he defined the solar rotation rate by watching the low-latitude sunspots. He also defined a solar reference system rotating with a period of 25.38 days. The first Carrington Rotation was described in November 9Th of 1853, when Carrington began his Greenwich photoheliographic series.

In stars the story is different given that the surfaces are not resolved. One of the most frequently used method for determining rotation period is, thus, measurements of stellar brightness variability. Interestingly, measurements of stellar brightness is one of the oldest tools of stellar astronomy. For example, already about 2nd century BC Hipparchus classified stars by their magnitude. As well the extensive compilation of glass plates collected by Henry Draper and his wife Anna Palmer Draper that given birth to the HD stellar identification catalog. As well, those photometric measurements gave a strong input that help opening the sky for the ladies of Harvard Observatory and their remarkable measurements of the stars during the late 1800's.

1.4   Space born photometry

1.4.1   Total Solar Irradiance

The integrated over all wavelengths total radiative flux from the Sun measured at one astronomical unit is called Total Solar Irradiance (TSI). Currently the established TSI value is 1361 W m^{-2}. Even though the TSI have been for a long time known as the solar constant

1 Introduction

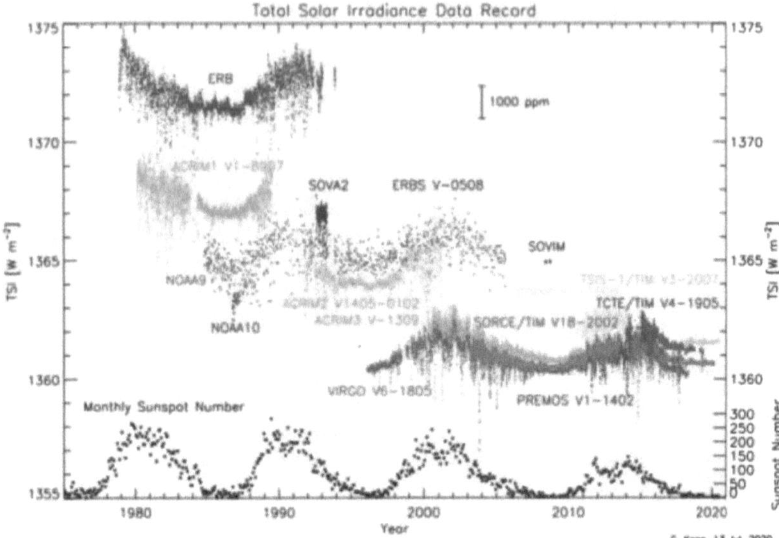

Figure 17: 41-years-long TSI dataset compilation by different instruments. Image taken

its value is not constant. The variations of the TSI have been reliable observed by precise radiometers during more than the 11-years of the solar activity cycle, and is about 0.1% or 1.3 Wm^{-2}. Over a solar rotation time scale the variation caused but large sunspots can be as high as 0.5% (see, Fröhlich 2013).

Driven by the interest from the climate community TSI has been measured almost without interruption for 41-year period by various satellite-based radiometers (such as, ACRIM, ERBS, VIRGO and TIM, see Fig. 17). Furthermore, it has been reconstructed for longer periods of time by many different models, eg. SATIRE and NRLTSI, (see for a review, Ermolli et al. 2013; Solanki et al. 2013; Ball et al. 2014; Dasi-Espuig et al. 2016; Yeo et al. 2014; Işık et al. 2018).

Among all available total solar irradiance records, the time series obtained by the Variability of solar IRradiance and Gravity Oscillations (VIRGO) experiment on the ESA/NASA SOlar and Heliospheric Observatory (SoHO) Mission and by the Total Irradiance Monitor (TIM) on board the Solar Radiation and Climate Experiment (SORCE, Dec. 2016 – 25 Feb. 2020) are the most accurate. They have also the longest time coverage, with 24- and 17- year-long data acquisition respectively (see e.g, the review by Kopp 2014). In this book I use these two TSI time series.

The first dataset used in this work was obtained by VIRGO/SoHO Mission, see Fröhlich et al. (1997). VIRGO provides more than 24 years of continuous high-precision, high-stability, and high-accuracy TSI measurements. Our analysis is based on the first 21 years of recorded data, and last update available at the beginning of this work, version 6.4: 6_005_1705, level 2.0 VIRGO/PMO6V observations from January 1996 until June

2017 with a cadence of 1 data-point per hour [2]. The data are available at the ftp server [3].

The second data-set used in this work was acquired by TIM/SORCE instrument (see, Kopp and Lawrence 2005; Kopp et al. 2005a,b). Regular TIM data used comes from version 17, level 3.0, with daily or 6-hourly cadence data. For our work we used data corresponding to: Feb 25th 2003 - Jan 25th 2018[4] [5]. While TIM data are available for a shorter time interval than VIRGO, they have lower noise level (Kopp 2014) which is particularly important for our analysis of TSI variations during the minimum of solar activity. Here we use an average version of TSI with a cadence of 1 data-point per 1.6 hours based on the regular data [6].

1.4.2   Stellar photometric data

The arrival of photometric planet-hunting missions, such as *CoRoT* (Bordé et al. 2003a), *Kepler* (Borucki et al. 2010), and *TESS* (Ricker et al. 2015), placed the studies of stellar magnetic activity to a completely new level. Continuously observations with high cadence are required to determine rotation period from stellar brightness variability. The missions mentioned have provided photometric time series with unprecedented precision and cadence. As a result, now it is possible to estimate rotation periods for thousands of stars. Stellar light-curves observed by *Kepler* have been employed in numerous studies aimed at determining rotation periods and stellar surface shear (for a complete description of different rotation period analysis methods frequently used see comparative study by Aigrain et al. 2015).

1.4.2.1   *Kepler* mission

The *Kepler* mission ran from 2009 to 2013. *Kepler* was designed with the scientific objective to explore the transits of the multiple possible Sun-like planetary systems in a region of the galaxy with a high density of solar-analogs expected, a region between the Cygnus and Lyra constellations (see, Borucki et al. 2010, and Fig. 1.8). Even though the primary goal of *Kepler* was planet-hunting, that made possible to revolutionise our understanding of stellar activity. The *Kepler* observations were in a band-pass covering from 420 to 880 nm. The instrument obtained and offered different data-products including, full-frame-images (FFIs), co-trending basis vectors, pixel response function, long and short cadence target light curves and pixel files, among others.

The data-product utilised in this work employed long cadence light curves (LCs) with a integrated 29.45 min cadence. The calibrated LCs and data-products are available at the MAST archive [7]. Those LCs were acquired on 17 segments called quarters (Q_1: (33 days), Q_2–Q_{16}: (90 days), and Q_{17}:(35 days), see public data release 25 and handbooks, Thompson et al. 2016; Van Cleve and Caldwell 2016; Bryson et al. 2017; Morris et al.

1   Introduction

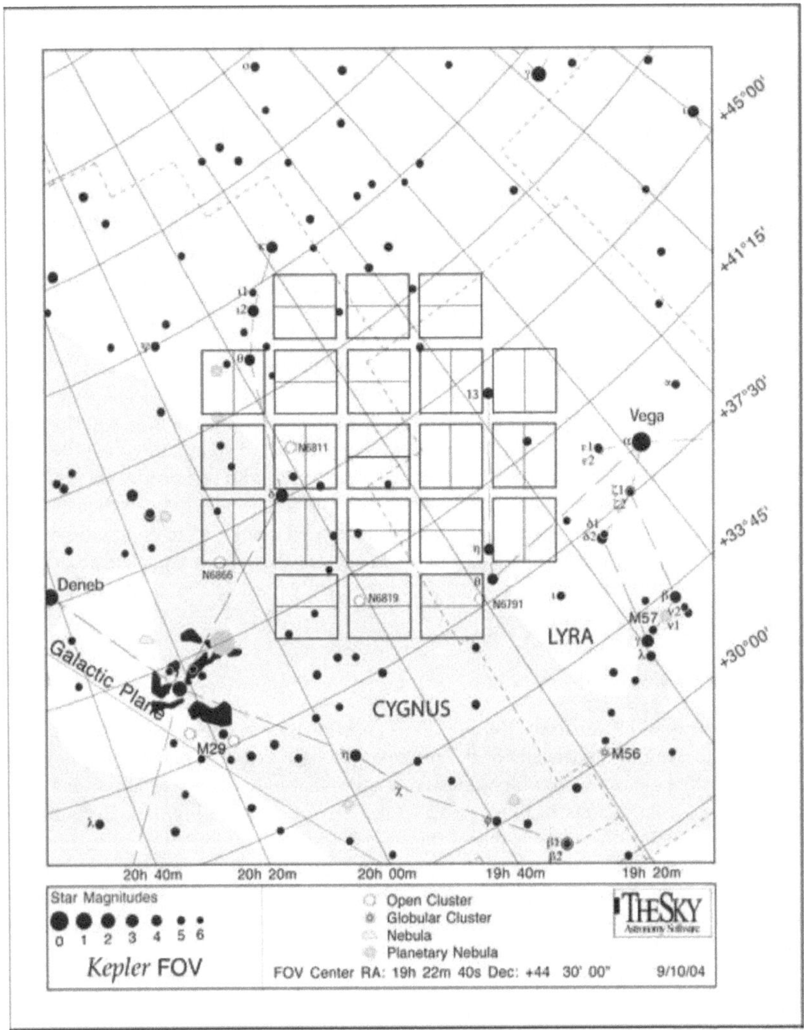

Figure 1.8: Field Of View of Kepler satellite, Credits: NASA Ames.

2017). The observed quarters are given in segments due to the Kepler telescope reoriented itself every 90 days. Quarter Q_1 is incomplete given the initial calibrations and Q_{17} due to a wheel failure that after 4-years of operation ended the original idea of the mission of continuous observation at the same FOV.

Kepler's field of view, FOV, has a resolution of 4 arc-seconds per pixel and contains approximately half-million stars. The satellite followed about 200,000 targets and around 150,000 stars were selected for continuous observation. More than 90,000 are G-type stars on, or near, the main sequence (see, Batalha et al. 2010). About 25% of stars from the primary *Kepler* FOV have reported rotation periods, (see, e.g., McQuillan et al. 2014, the largest rotation period survey presently available). Interestingly, it implies that we do not know rotation periods of almost 75% of G-type stars. In particular, we lack information about rotation periods in low-activity stars like the Sun. The biggest difficulties for determining rotation periods of such stars from photometric records are associated to non-periodic light-curve profiles and low amplitude of the variability.

1.4.3   Rotation period in planetary transits analysis

Although that the goal of this book is not connected with planetary transit analysis, the knowledge of precise rotational periods are required for removing stellar activity signal present in the light-curves. Stellar activity can mimic planets in radial velocity analysis (RV) as well as affect characterisation of both RV and transiting planets (see, Fig. 1.9). Knowing precisely and accurately the stellar rotation allow disentangling the signal from star and planet in RV. It, in turn, can help to detect small-sized planets in RV, which is crucial for ongoing and upcoming survey like ESPRESSO. For transiting planet it will also help to get more accurate planet radius estimation. The characterisation of exoplanets have been improved using simulations of spots to correct transit events, but still the disambiguation from activity needs to be more explored (see, Dumusque et al. 2011; Oshagh 2018). Rotation period acquaintance is decisive to constrain models on activity-transit entangle and star-planet interaction.

1.5   Photometric methods for rotation periods detection

Thanks to planetary hunting missions such as CoRoT, *Kepler* and TESS the possibilities of acquiring accurate photometric time series with high resolution and high cadence are now real. Building on those high quality observations, the scientific community has developed different methods and techniques to analyse and interpret stellar periodicities embedded in the data. Some of the current methods include autocorrelation functions analysis, Lomb-Scargle periodogram, periods based on wavelet power spectrum, and recently techniques based on Gaussian processes.

1.5.1   Generalized Lomb-Scargle periodogram – (GLS)

The Lomb-Scargle periodogram (hereafter, GLS) is a formalism used to analyse the frequency domain of unequally space time-series. It is analogue to fitting a sinusoidal function, $y = a\cos\omega t + b\sin\omega t$. The first formalism was given by Barning (1963) and

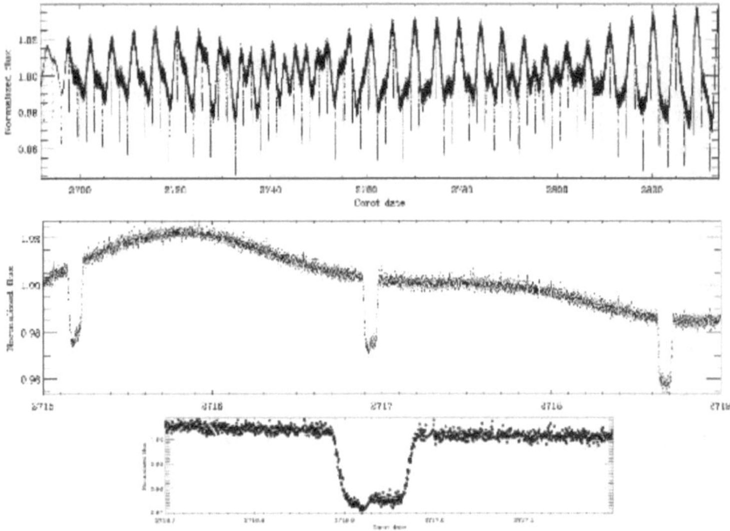

Figure 1.9: Conjugation of rotation period modulation (top) and planetary transits (middle and bottom) in CoRot-Exo-2b system, (taken from, Alonso et al. 2009).

afterwards Lomb (1976); Scargle (1982) analysed the statistical significance of a periodic signal.

In the original formalism the measurement of the errors are not considered and, it is assumed that the mean of the data and the mean of the fitted function are the same. For an improved analysis in this work I consider the Generalised Lomb-Scargle periodogram (GLS) version v1.03, applying the formalism given in Zechmeister and Kürster (2009).

The GLS method is widely used for time domain analysis and has the advantage for treating data-sets with a non-regular sampling. For rotation period detection purposes, the highest normalised power peak is usually assumed to correspond to the rotational period (see GLS applied to stellar LCs in, Reinhold et al. 2013; Aigrain et al. 2015; Reinhold et al. 2019, 2020b). An example of the computed GLS for a TESS light-curve of the target TIC 441420236 is shown in the panel b of Fig. 1.10.

1.5.2   Auto-Correlation Functions – (ACF)

Autocorrelation functions (hereafter, ACF) is a method based on the estimation of a degree of self-similarity in the light-curve over time. The time lags at which the degree of self-similarity peaks are assumed to correspond to the stellar rotation period and its integer multiplets.

The ACF method was introduced as a statistical model for exploratory data analysis (EDA) initially implemented for climatology and hydrology time-series and has been widely used in many fields since then (see, Yevjevich 1968; Merz et al. 1972). ACF

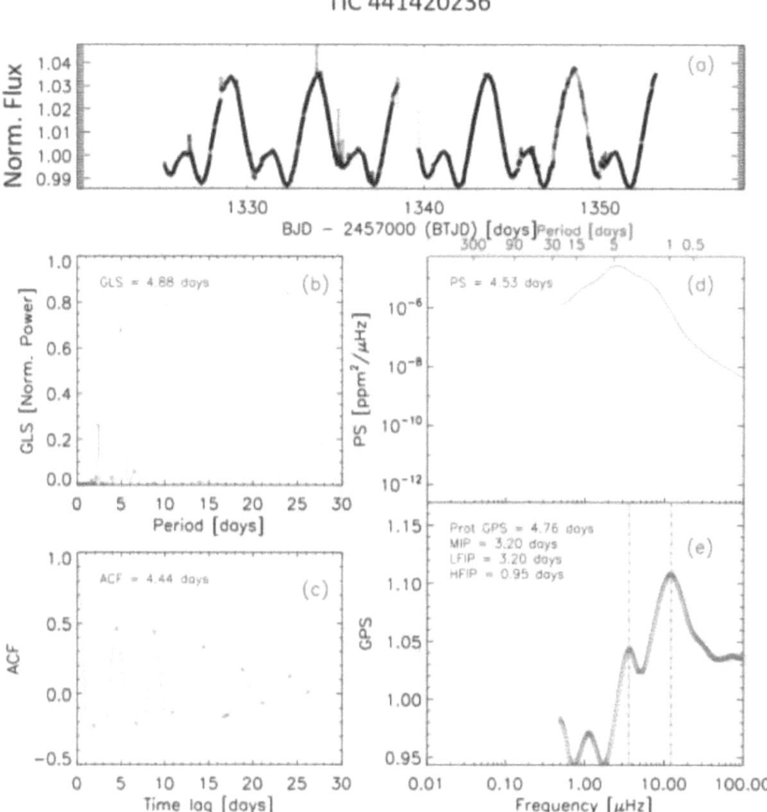

Figure 1.10: Compilation of rotational period outputs for the object TIC 441420236 computed with GLS, ACF, PS, and GPS. Panels (a) show the TESS LC, (b) the GLS output, (c) The ACF, (d) the wavelet power spectra using Paul wavelet order 6, and (d) the GPS outcome.

have the main objective to detect the non-randomness in the data. It characterises the self-similarity in the given measurements, $Y_1, Y_2, ..., Y_N$ at the times $X_1, X_2, ..., X_N$ as a function of the lag k:

$$r_k = \frac{\sum_{i=1}^{N-k}(Y_i - \overline{Y})(Y_{i+k} - \overline{Y})}{\sum_{i=1}^{N}(Y_i - \overline{Y})^2} \tag{1.1}$$

The time variable, X, is not used in the formula, but the assumption is that the observations are equally spaced. Usually the first maximum of the autocorrelation is taken as the searched periodicity.

The ACF method applied for the analysis of stellar time series was introduced by

1   Introduction

LC Number	A_CORRELATE	AutoACF
1000	26.58	26.71
1001	21.07	21.24
1002	10.29	10.39
1003	27.07	27.04
1004	26.21	26.06

Table 1.1: Comparison of solar rotation periods output from the IDL A_CORRELATE routine and the AutoACF for the 5 different solar light-curves proposed in Aigrain et al. (2015)

McQuillan et al. (2013). The application of the ACF to stellar LCs is based on the assumption that magnetic features which cause photometric variability are stable over the stellar rotational period. The ACF has been used to create the largest available catalog of rotational periods until now. Using ACF valuable statistical information for about 34000 stars observed by *Kepler* have been compiled and analysed in detail in McQuillan et al. (2013); McQuillan et al. (2014).

In the present work I tested and implemented the autocorrelation function approach given by the IDL A_CORRELATE [8] routine comparing with the results given by the AutoACF method, introduced in McQuillan et al. (2013), and used by the Tel Aviv team in the hare-and-hounds exercise in Aigrain et al. (2015). They, performed a blind exercise to compare different methods to obtain rotation periods from 1000 simulated light-curves injected to 770 Kepler and 5 solar SoHO/VIRGO light-curves. We calculate the autocorrelation function from A_CORRELATE and compare with the outputs from AutoACF for the 5 solar light-curves for different ranges of solar activity. We show the comparison of both algorithms in table 1.5.2, and Figure 3.1. In the AutoACF implementation, the light curves are median normalized before the ACF is computed, and they only search for periods less than half the length of the data set. We verified that the outputs from the ACF IDL A_CORRELATE routine and AutoACF have a similar behaviour. An example of the computed ACF for a TESS light-curve of the target TIC 441420236 is show in the panel c of Fig. 1.10.

1.5.3   Wavelet Power Spectra – (PS)

Wavelet power spectra analysis (hereafter, PS) is beneficial for time series that contain non-stationary power at many different frequencies. PS was originally used to analyse geophysics and climatology time-series. Recently, in combination with the ACF it has been also employed for determining stellar rotational periods (see, García et al. 2009; Aigrain et al. 2015; Santos et al. 2019). An important aspect of the PS method is the choice of the wavelet function, $\Psi(\eta)$. There are many different wavelet functions, Morlet, DOG, Paul, etc (see, Fig.1.12 and Torrence and Compo 1998). To calculate the PS in this work I used the WV_CWT [9] IDL function. It is based on Paul wavelet of order $m = 6$. Paul wavelet is a complex non-orthogonal function, which means that the wavelet will return information

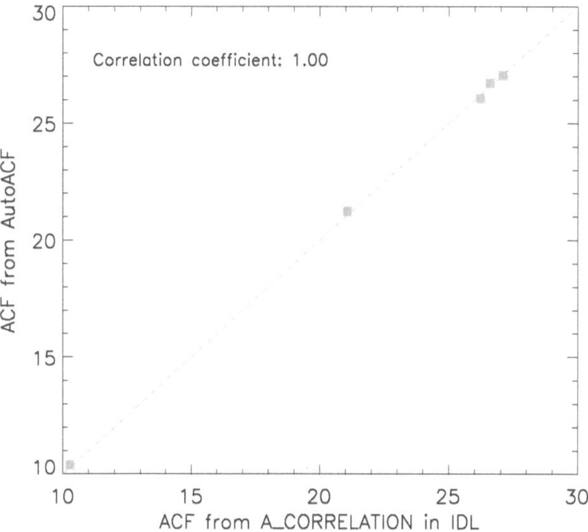

Figure 1.11: Comparison between ACF from AutoACF and the IDL A_CORRELATE routine.

about the amplitude and, if required, phase of the signal. An example of the computed PS for a TESS light-curve of the target TIC 441420236 is show in the panel d of Fig. 1.10.

1.5.4   Gaussian Process – (GP)

Gaussian processes are certainly an area of extremely active investigation in astrophysics at the moment. Statistical Gaussian processes (hereafter, GP) can be applied to detect a non-sinusoidal and quasi-periodic behaviour of the signal in light-curves. Since recently, the GP have being used for retrieving the periodic modulations from stellar activity (see Roberts et al. 2012; Rajpaul et al. 2015; Angus et al. 2018). The GP will fit a good inference for periodic patterns for a singular light curve, but will take several hours to converge, (see, e.g. Angus et al. 2018). For optimizing the procedure Angus et al. (2018) sub-sampled Kepler light-curves by a factor of 30 and split them into segments of 300 points. To improve considerably the computationally efficiency, Foreman-Mackey et al. (2017b,a) developed a new algorithm which scales linearly with the number of data N points instead of $NLog(N)^2$.

Even with a linear scaling GP calculations demand significant computational resources. Such methods can be extensively implemented and compared with other approaches for determining rotation periods in a limited number of stars, as for example in the analysis performed for HD 41284 in (Faria et al. 2020, see, Appendix 1.8). GP method is out of

Name	$\psi_0(\eta)$	$\hat{\psi}_0(s\omega)$	e-folding time τ_s	Fourier wavelength λ
Morlet (ω_0 = frequency)	$\pi^{-1/4} e^{i\omega_0 \eta} e^{-\eta^2/2}$	$\pi^{-1/4} H(\omega) e^{-(s\omega - \omega_0)^2/2}$	$\sqrt{2}s$	$\dfrac{4\pi s}{\omega_0 + \sqrt{2 + \omega_0^2}}$
Paul (m = order)	$\dfrac{2^m i^m m!}{\sqrt{\pi(2m)!}}(1 - i\eta)^{-(m+1)}$	$\dfrac{2^m}{\sqrt{m(2m-1)!}} H(\omega)(s\omega)^m e^{-s\omega}$	$s/\sqrt{2}$	$\dfrac{4\pi s}{2m+1}$
DOG (m = derivative)	$\dfrac{(-1)^{m+1}}{\sqrt{\Gamma\left(m+\frac{1}{2}\right)}} \dfrac{d^m}{d\eta^m}\left(e^{-\eta^2/2}\right)$	$\dfrac{i^m}{\sqrt{\Gamma\left(m+\frac{1}{2}\right)}}(s\omega)^m e^{-(s\omega)^2/2}$	$\sqrt{2}s$	$\dfrac{2\pi s}{\sqrt{m+\frac{1}{2}}}$

$H(\omega)$ = Heaviside step function, $H(\omega) = 1$ if $\omega > 0$, $H(\omega) = 0$ otherwise.
DOG = derivative of a Gaussian; $m = 2$ is the Marr or Mexican hat wavelet.

Figure 1.12: Comparison between three wavelet functions and its properties, (taken from Torrence and Compo 1998).

the scope of this work.

1.5.5   Gradient of the Power Spectra: GPS

In this book I has been developing a new method for the determination of stellar rotation period. The method is based on the analysis of the gradient of the power spectra (GPS) of stellar brightness variations. In contrast to the methods described before GPS method is aimed at low-activity stars like the Sun, but also works in more active stars. In Chapter 2 of this book we develop a mathematical formulation of the method. In particular, we show that the profile of the power spectrum around rotational period depends strongly on the decay time of active regions. It is also possible that the rotation peak absent from the power spectrum at all. For example, it will be flatted or absent for low activity stars, like the Sun, which rotation period is longer than the decay time of magnetic features. Furthermore, there could be rogue peaks which do not correspond to the rotation period but could be easily misinterpreted with the rotation peak. Despite this we show that the profile of the high-frequency tail of the power spectrum remains stable and only weekly depends on the evolution of magnetic features. This allows us to propose using inflection point, i.e. the point where the concavity of the power spectrum plotted in the log–log scale changes its sign, as a sensitive diagnostic of the rotation period.

Chapter 3 contains the application of the GPS method to TSI observations. The method is compared with regular methods used for detecting rotation periods on stars. GPS recover more accurate values for the solar rotation period in the comparison with the other methods, independently of the activity regime of the star. We also show that GPS can be used to distinguish periods of facular or spot dominance in the stellar brightness variation.

After successfully testing GPS against simulated and observed solar light-curves GPS is applied in Chapter 4 to brightness time series of observed stars in the *Kepler* field. For testing GPS on stars, were selected targets with know rotation periods reported by Reinhold et al. (2013); McQuillan et al. (2014). The light curves of the samples presented a regular modulation and higher variability than the solar TSI. That characteristics allowed to all the methods applied to easy recover the rotation period on *Kepler* stars. GPS rotation period values are well correlated with the previously reported values by other methods. In that way GPS is tested and verified.

Furthermore, GPS can be used to estimate the faculae to spot area ratio (S_{fac}/S_{spot}). In Chapter 4 the S_{fac}/S_{spot} is calculated for a stellar sample. We show that facular to spot ratio decreases with the increase of stellar ration rate.

Additionally to *Kepler* light-curves, GPS was applied on TESS light-curves. An example of the computed GPS for a TESS light-curve of the target TIC 441420236 is show in the panel e of Fig. 1.10.

1.6   State of the Art

Rotation periods are not equally detectable for all stellar objects. Observational and theoretical studies as in Aigrain et al. (2015); van Saders et al. (2019) show that rotation periods in cool dwarfs around solar effective temperature and with a low magnetic activity are more difficult to detect using current methods. The estimation of rotation period on stars with a similar activity behaviour than our Sun is difficult, even with the advantage given by high quality data from space born missions.

In He et al. (2015, 2018) they analysed the solar and stellar activity using GLS and introducing two indicators, one by describing the degree of periodicity on the light-curve, i_{AC}, and the other by the effective fluctuation range, R_{eff}, that describes the deep of the rotation modulation. They found that light-curves periodicities of the *Kepler* stars were generally stronger in maximum season of activity than the one of the Sun, where the highest periodicity was determined during low active seasons of activity. By applying GLS and the indicators to the TSI they identify the solar rotation period only during solar activity minimum regime. A similar result was found by Aigrain et al. (2015), where they compare, in a blind exercise some of the methods mentioned as, ACF, GLS and PS to retrieve rotation periods from simulated light-curves and real data from the Sun. In addition, they reported that the rotation period values were not equally detected for all methods, probably due the different levels of signal/noise in the simulated light-curves.

In van Saders et al. (2019) they presented a theoretical approach that assume a relationship between Rossby number with the amplitude of the variability. They estimate a threshold in the Rossby number that can characterise the level of detectability of stellar rotation by spot-modulation. Those threshold values could be representative of the level of activity, assuming the relationship between R_o with the amplitude of the variability and that they describe the level of detectability of rotation by modulation in cool dwarf stars. Their models suggested that exist a limit in the spot modulation amplitude below which period detection from current methods is inefficient. They reproduced Kepler-like observational

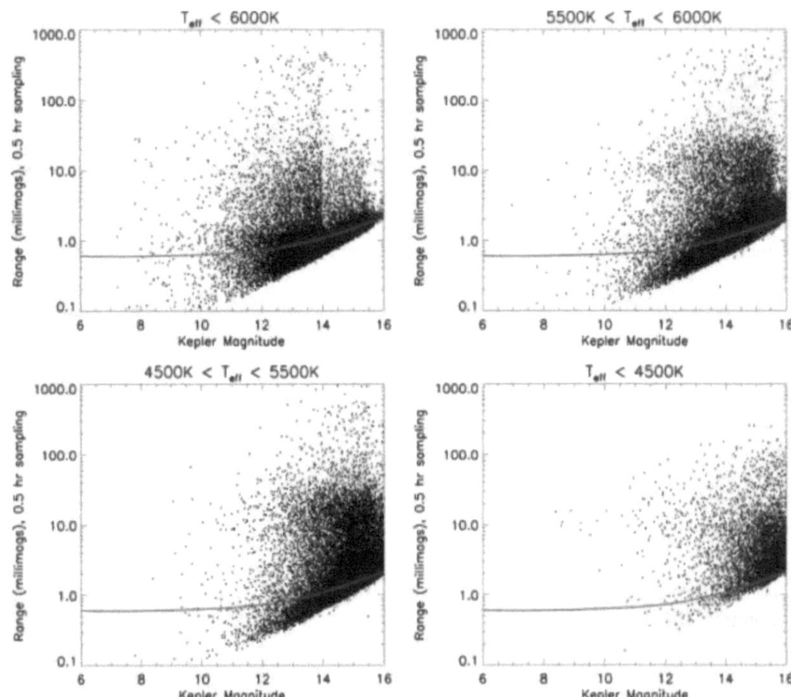

Figure 1.13: *Kepler* magnitude vs. the range of light curve variability in different temperature ranges. The active Sun is shown with a red line (see, Basri et al. 2010).

bias on models and show that highest rotation detection fraction discrepancies are around solar temperature, at $T_{eff} \approx 5700$ K, (see figure 13 in, van Saders et al. 2019). They confirmed that 80% of stellar rotation periods in the *Kepler* field of view with near-solar effective temperature remain undetected.

For the solar case, where $R_0 = 2.01$, Brandenburg and Giampapa (2018) proposed that for $R_o \gtrsim R_{o\odot}$ there are two possible scenarios: one where stars that reach solar Rossby number start a process to reduces its magnetic braking and then become less active, or two, that stars enter in a regime of anti-solar differential rotation, in other words where poles rotate faster than the equator (see, Viviani et al. 2018, 2019). In (Basri et al. 2010, 2013) solar variability appears to be normal when compared to main-sequence *Kepler* stars with near-solar effective temperatures, see Fig. 1.13. Even though the amount of reported rotation periods of stars with near solar variability and parameters are lower than expected in the *Kepler* field. The current knowledge of rotation periods is restricted to stars with strong variability and regular modulation, to more active cool stars unlike the Sun, at least in terms of its variability and magnetic activity. Due to the detectability difficulties described before, the information of rotation periods of solar analogs available in the literature is just the peak of the iceberg, it telling us that only a small fraction of

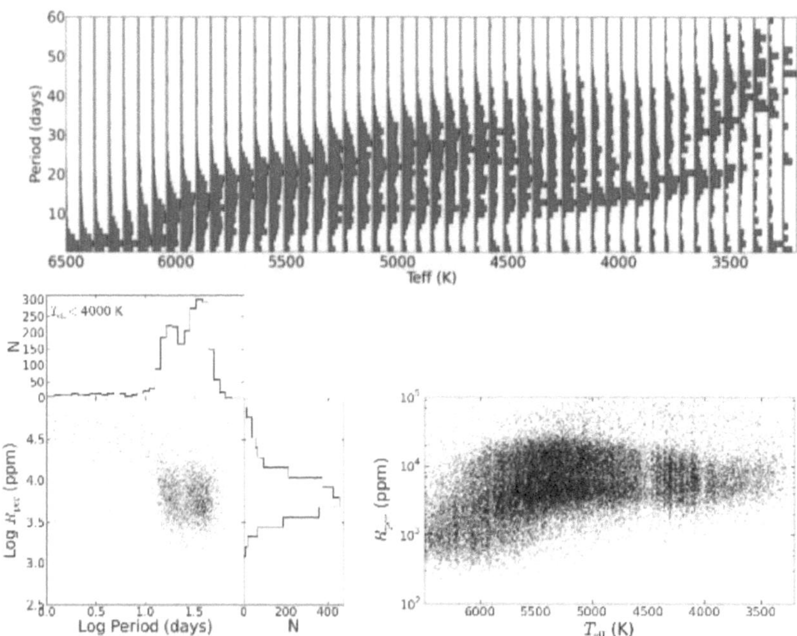

Figure 1.14: Top: Rotation periods detected by autocorrelation function method versus effective temperature. It shows bimodal rotation period distribution for different ranges of temperature. Bottom: Left: Amplitude versus rotation period for M-dwarfs with T_{eff} about 4000 K. Is observed a decrement of stars with rotation period near 19-21 days and visible a bimodal distribution. Right: Amplitude versus effective temperature for M-dwarfs stars in the *Kepler* field (For more details see, McQuillan et al. 2013).

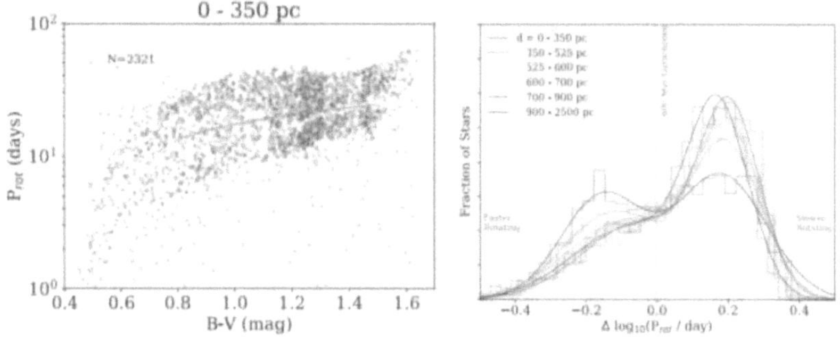

Figure 1.15: Bimodal rotation period distribution for *Kepler* stars from 0 to 350 pc (For more details see, Davenport and Covey 2018).

solar-like systems have been analysed. Shapiro et al. (2016) demonstrated that the main reason for the irregular temporal profile of solar variability is attributed to the short-time evolution of sunspots. In other words just few magnetic features last longer enough to reproduce the sinusoidal signal from the rotation. Furthermore, Shapiro et al. (2017) show that facular and spot contributions to the solar brightens cancel each other signal on the power spectrum over the rotation period time-scale.

In Metcalfe et al. (2016) and Metcalfe and van Saders (2017) proposed that the Sun could be in a transition state to a different low-activity dynamo regime, and stars with a clear periodicity are still in a high-activity regime. Now, Reinhold et al. (2020a) show that the solar variability appears to be anomalously low when is compared with main-sequence stars with near-solar effective temperature and with *known* near-solar rotation periods. An additional picture to explanation such a paradox is the inability of standard methods to reliably detect rotation periods of stars with variability similar to that of the Sun (see also discussion in, Witzke et al. 2020). The outcome of GPS could bring us the tool to analyse and understand such a paradox. In Reinhold et al. (2019) suggested that biases in determining rotation periods might contribute to the explanation of a dearth of intermediate rotation periods observed in *Kepler* stars (see McQuillan et al. 2013; Reinhold and Gizon 2015; McQuillan et al. 2014; Davenport 2017; Davenport and Covey 2018). In other words, long period with low amplitude stars are hardest to detect as shown in Figs. 1.14 and 1.15.

In (Shapiro et al. 2020; Amazo-Gómez et al. 2020b) we showed that the rotation periods of the Sun from observed total solar irradiance (TSI) and simulated lightcurves of closer stellar analogs with similar solar fundamental parameters, can be reliably determined from the profile of the gradient of the power spectrum, GPS. We retrieve rotational period values base on the automated GPS method, a novel rotational analysis method that follows the characteristics imprinted by spots and faculae on the gradient of the power spectrum.

Differences in the CLV contrast reflected in the light-curves for spots and facular regions are the starting point of GPS. After analysing that the contribution from faculae and spots have differences in the power spectra profile related with its relative V-like and M-like light-curve shapes (see, Fig 1.5), we were able not just to determine rotation period but facular or spot dominance in the solar surface. The manifestations of facular- and spot-related signatures respectively on the third and second harmonic of the rotation period value can be characterised by the inflection points at the GPS.

The summarized ideas previously introduced in the state of the art along to a detailed explanation of the GPS method will be expanded throughout the following chapters in this book.

GPS applied on collaboration papers:

The following sections of the Introduction includes the abstract and the contribution to three different publications in which I participated as co-author. I implemented the GPS method and/or contributed with ideas and analysis in the context of the present book.

1.7   The Sun is less active than other solar-like stars

This section is based on the article published at Science Journal, volume 368, pages 518-521, by Timo Reinhold, Alexander I. Shapiro, Sami K. Solanki, Benjamin T. Montet, Natalie A. Krivova, Robert H.Cameron & Eliana M. Amazo-Gómez. I contributed in this manuscript analysing differences between Solar and Stellar variability. The printed version is reproduced here with permission from Science Journal, © AAAS.

Abstract SM-A

Magnetic activity of the Sun and other stars causes their brightness to vary. We investigate how typical the Sun's variability is compared to other solar-like stars, i.e. those with near-solar effective temperatures and rotation periods. By combining four years of photometric observations from the *Kepler* space telescope with astrometric data from the Gaia spacecraft, we measure photometric variabilities of solar-like stars. Most of the solar-like stars with well-determined rotation periods show higher variability than the Sun and are therefore considerably more active. These stars appear nearly identical to the Sun, except for their higher variability. Their existence raises the question of whether the Sun can also experience epochs of such high variability.

1.7.1   Rvar distribution

Figure 1.16 shows the distribution of R_{var} for the Sun, the periodic stars, and a composite sample of the periodic and non-periodic samples combined. To compare the Sun with the stars observed by Kepler, we simulated how it would have appeared in the *Kepler* data by adding noise to the TSI time series. The variability range was then computed for 10,000 randomly selected 4-year segments from ~140 years of reconstructed TSI data.

Figure 1.16: **Solar and stellar variability distributions on a logarithmic scale.** The distributions of the variability range R_{var} are plotted for the composite sample (black), the periodic sample (blue), and the Sun over the last 140 years (green).

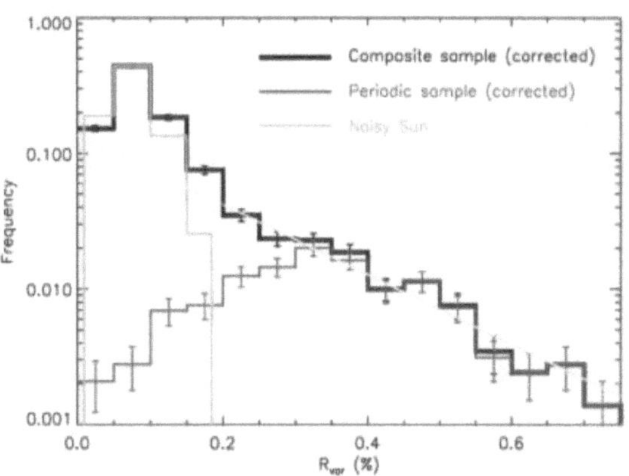

1.8   Decoding the radial velocity variations of HD41248 with ESPRESSO

1.8.1   TESS

The Transiting Exoplanet Survey Satellite (TESS; Ricker et al. 2014, 2015) observes HD41248 (TIC 350844714, TESS magnitude = 8.187) in sectors one through 13 of its nominal two-year mission. As of June 2018, data from the first ten sectors are available (from 25 July 2018 to 22 April 2019). This leads to a baseline of around 270 days. TESS observations are simultaneous with the ESPRESSO RVs between the end of sector four and middle of sector nine.

We downloaded, combined, and analysed the TESS light curves for the first 10 sectors. An in-depth analysis of the combined light curve is described in 1.8.2. In summary, we do not detect credible transit signals. We do find evidence for a stellar rotation period between 24 and 25 days. The data are consistent with a spot lifetime of about 25 days.

1.8.2   Analysis of the TESS light curve

The TESS mission is set to observe HD41248 during the full first year of its nominal two-year mission. Using the Lightkurve package (Lightkurve Collaboration et al. 2018), we downloaded and extracted the Pre-search Data Conditioning (PDCSAP_FLUX) light curves (LC) produced by the Science Processing Operations Center from the Mikulski Archive for Space Telescopes (MAST [10]). As of June 2018, data from the first ten sectors are available, with a baseline of 243 days. The individual LCs were then merged by adjusting the mean of the flux in each sector, and outliers were removed with a 5-sigma-clipping procedure. This results in the merged LC shown in Fig.1.17, which also includes an indication of the period where TESS observations are simultaneous with ESPRESSO.

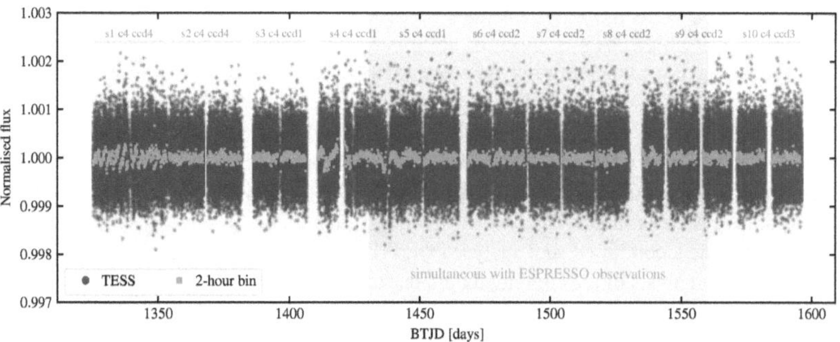

Figure 1.17: Merged TESS light curve from the first ten sectors. The camera and CCD number with which HD41248 was observed in each sector is indicated at the top, as well as the period of ESPRESSO observations. The orange points show the binned LC over a 2-hour window.

The merged LC shows a weighted rms of 431 ppm. Using the relations between active-region lifetime, spot size, and stellar effective temperature determined by Giles et al. (2017, their Eq. 8). This leads to an estimate of 25.57 days for the decay lifetime of active regions in the stellar surface. This relation was built for star spots, since these have a larger effect in the brightness variations when compared with faculae. In the Sun, faculae tend to live longer than spots (Solanki 2003; Shapiro et al. 2017).

1.8.3   Rotation period

We searched the TESS LC for a periodic signal that can be associated to stellar rotation using four different methods: the GLS periodogram, the autocorrelation function (ACF, e.g. McQuillan et al. 2014), the wavelet power spectra (PS, e.g. Torrence and Compo 1998),

---

[10] mast.stsci.edu/portal/Mashup/Clients/Mast/Portal

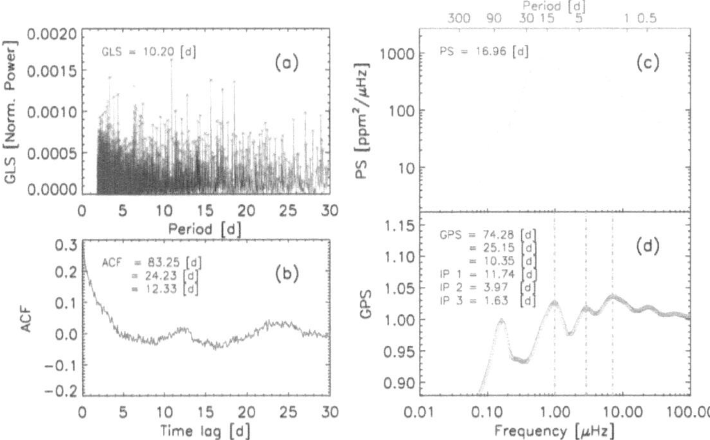

Figure 1.18: Results from the rotation period analysis showing the GLS periodogram (panel a), ACF (panel b), power spectrum (panel c), and GPS (panel d) of the TESS LC. Each panel displays the most prominent periods detected with each method.

and the gradient of the power spectra (GPS[11], Shapiro et al. 2020, Amazo-Gómez et al. 2020ba, and Amazo-Gómez et al. 2020ab)

The GPS method in particular attempts to determine the rotation period from the enhanced profile of the high-frequency tail of the power spectrum by identifying the point where the gradient of the power spectrum reaches its maximum value. Such a point corresponds to the inflection point (IP), that is, a point where the concavity of the power spectrum changes sign. Shapiro et al. (2020) show that the period corresponding to the inflection point is connected to the stellar rotation period by a calibration factor equal to $\alpha_{Sun} = 0.158$, for Sun-like stars.

The results from the four methods are presented in Fig. 1.18 and can be summarised as follows: the GLS periodogram suggests a periodic signal of 10.2 days, but with a low relative power; the ACF shows periodic signals at 24.25 days and 12.34 days. The PS, in panel (c), shows two peaks at 16.96 days and 6.15 days. The GPS method shows three enhanced inflection points with enough amplitude to determine three different periodicities. The inflection points at 11.74, 3.97, and 1.63 days correspond to periodic signals at 74.28, 25.15, and 10.35 days after applying the calibration factor α_{Sun}.

From the values obtained using the four different methods, we can see that both the GLS and GPS methods detect a periodicity close to 10 days. The strongest signal in the ACF is around 24.25 days, in agreement with the second enhanced signal from GPS, of 25.15 days. The values obtained with the ACF and GPS are close to those obtained from spectroscopy ($v \sin i$) and with the periodicities seen in some activity indicators, suggesting a stellar rotation period for HD41248 of about 25 days.

---

[11] Not to be confused with Gaussian processes, GPs.

1.9   The correlation between photometric variability and radial velocity jitter

1.9.1   Light curve: TESS

We obtained the light-curves of all 171 stars from The Mikulski Archive for Space Telescopes (MAST). MAST contains TESS simple aperture photometry (SAP_flux) (Morris et al. 2017) as well as presearch data conditioning (PDCSAP_flux). Most of the targets (96 %) were observed in only during one TESS sector. For stars with light-curves in two or more consecutive sectors, we merged all available light-curves. We

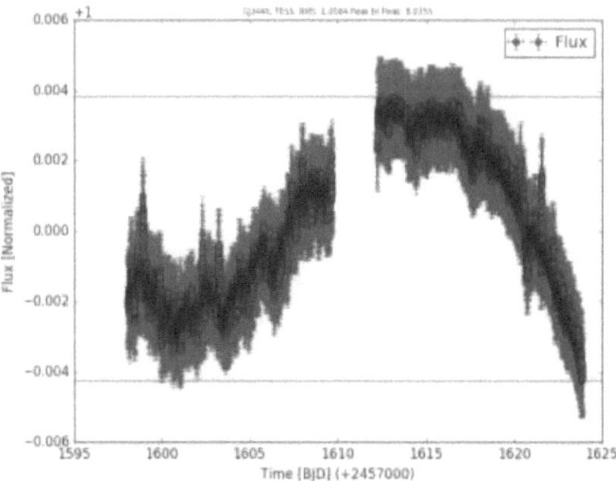

Figure 1.19: Light curve of **GJ3440** during one TESS sector (blue points) and the red lines presenting the peak-to-peak of light curve.

used quality-flag[12] as suggested by the TESS Data Product review, also recently used and tested in (Vida et al. 2019). We used SAP_flux which optimizes the aperture for the best signal-to-noise for the each target (Bryson et al. 2017) and also the calibrated pixels in order to perform a summation of the flux. The SAP light curves provided by the TESS pipeline are also background corrected. We removed outlier flux points using a sigma-clipping of three standard deviations and smoothed the fluxes using Savitzky-Golay filter within 15 data point windows ($\sim$ 30 minutes) to reduce the effect of the short-time scale photometric variability. Since we are interested in variability at stellar rotation timescales, this smoothing does not affect our results. We then normalized the flux by the median flux values. We derived the ratio between the peak-to-peak of light curve variability of SAP_flux and the peak-to-peak of light curve variability of PDCSAP_flux. If this ratio was larger than 3.0, we checked the light curves visually to ensure if there was any evidence for systematic errors in SAP_flux. For more than 90 % of the stars, we used SAP_flux, and for the rest we assumed that the light-curves are dominated by systematic errors therefore we used PDCSAP_FLUX. In Fig. 1.19, you can see a sample of reduced light curve in one sector for one star.

1.9.2   Stellar rotation period

Photometric contrast differences associated to magnetic features (e.g., dark spots and bright faculae) generate traceable signatures of stellar rotation periods on light curves.

We analyze the presence of a periodic modulation signal from stellar rotation on the TESS photometric time-series using the gradient of the power spectra (GPS) (see Shapiro et al. 2020; Amazo-Gómez et al. 2020b,a). We successfully recover the rotation period for 71 out of 171 stars of the sample. We report the estimated rotation period from the GPS

---

[12] 101010111111:
```
Overview
```

method in Hojjatpanah et al. (2020). The rotation period from GPS is determined from the enhanced profile of the high-frequency tail of the power spectrum. In particular, we identify the point where the gradient of the power spectrum GPS in log-log scale reaches its maximum value. Such a point corresponds to the high frequency inflection point (HFIP), that is, where the concavity of the power spectrum plotted in the log-log scale changes sign. The position of inflection point is related to the rotation period of star by the calibration factor α_{Sun}, for Sun-like stars.

For the calculations presented in this project we adopt a solar-like calibration factor $\alpha_{Sun} \pm 2\sigma = 0.158 \pm 0.014$, and 2 sigma uncertainty (for more details see, Shapiro et al. 2020; Amazo-Gómez et al. 2020b,a, b).

We also estimated the faculae to spot ratio for 29 of the 71 stars. Following the GPS outcome, we applied the criteria indicating that the light curve is faculae dominated when the ratio between HFIP and the independent rotation period ranges between [0.11-0.16], and spot is dominated when the value falls between [0.16-1.24].

1.9.3   Correlation with stellar rotation period

In Top panel in Fig. 1.20, we present RV-RMS and the peak-to-peak light curve variation for the subsample of 71 stars with measured rotation periods (coded with marker size). The color bar represents the effective temperature. One can easily notice that stars with large RV-RMS and a large peak-to-peak photometric variability are mostly fast rotating stars (less than 13 days) and there is a hint of temperature dependency. Bottom panel in Fig. 1.20 shows the same 71 stars but color bar indicates the rotation period value, and circle size the *v sin i* obtained spectroscopically, which again confirms the previous result we found.

In Fig. 1.21, we present a similar plot to Fig. 1.20 but this time for 29 stars where we could identify facular or spot dominated patterns using the method described in Sec. 1.9.2. We found 9 faculae dominated stars (which were also slow rotators as is expected for faculae dominated stars), and 20 spot-dominated stars. We show faculae and spot dominated stars in yellow and black, respectively. In this sample, 20 stars can be classified as fast rotators (rotation period < 15 days), and large fraction of them (13 out 20) are spot dominated. This result is in strong agreement with Montet et al. (2017), where they reported 15 days as the threshold in rotation period for separating spot-faculae dominated regimes. Moreover, faculae-dominated stars tend to have low photometric peak-to-peak variability, due to the low contrast of facular region, and therefore are mostly below the 6.5 ppt limit. Thus, the 6.5 ppt limit can be also interpreted as the photometric variability transition between the spot-dominated and the faculae-dominated regime. However, the sample of 29 stars is too small to generalize.

We were able to estimate the rotation period of 71 stars, out of 171 stars in our sample, using the TESS light curve. Then we investigated the effect of this parameter on the correlation between RV-RMS and peak-to-peak of light curve variability. Our result demonstrated that slow rotating stars (which are the ones also we found to be faculae dominated) create lower RV jitter as well as lower peak-to-peak photometric variability, and on the other hand fast rotating star, $P_{rot} \leq 5$ day (which are the ones also we found to be spot dominated) generate much larger RV jitter and photometric variability.

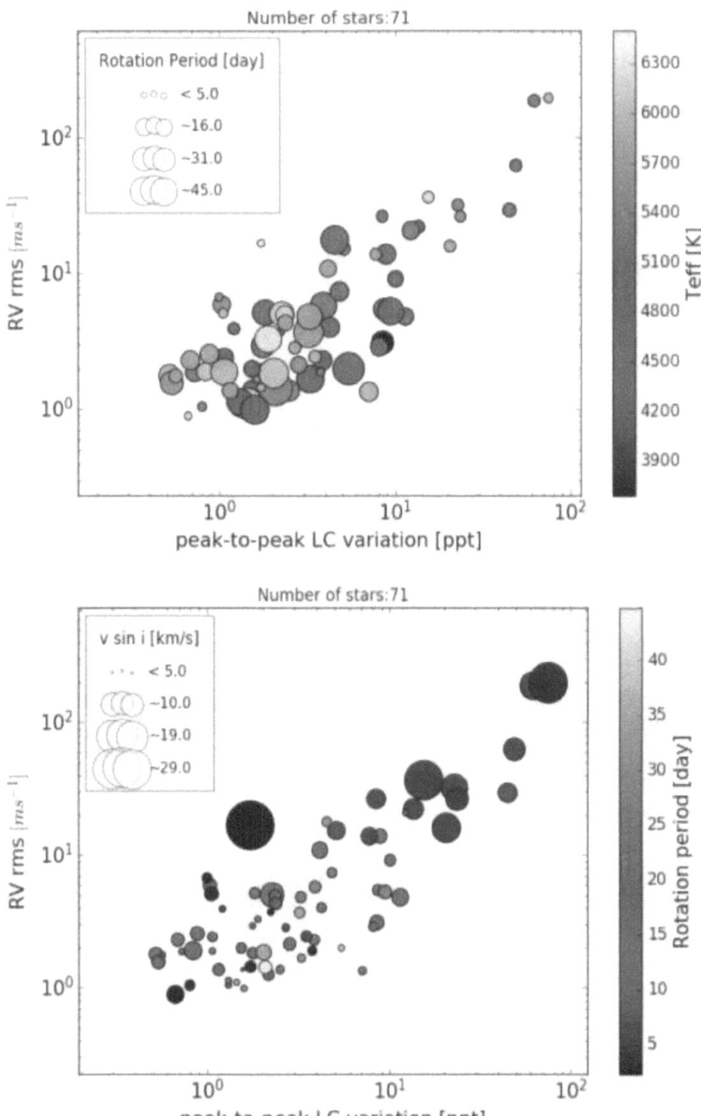

Figure 1.20: RV-RMS and peak-to-peak of light curve variation for the subsample of 71 stars. Top plot: Circle sizes represents the value of rotation period found by GPS method. Color bar indicates the stellar effective temperature. Bottom plot: Similar than top panel but, color bar indicates the rotation period value, and circle size the *v sin i* obtained spectroscopically

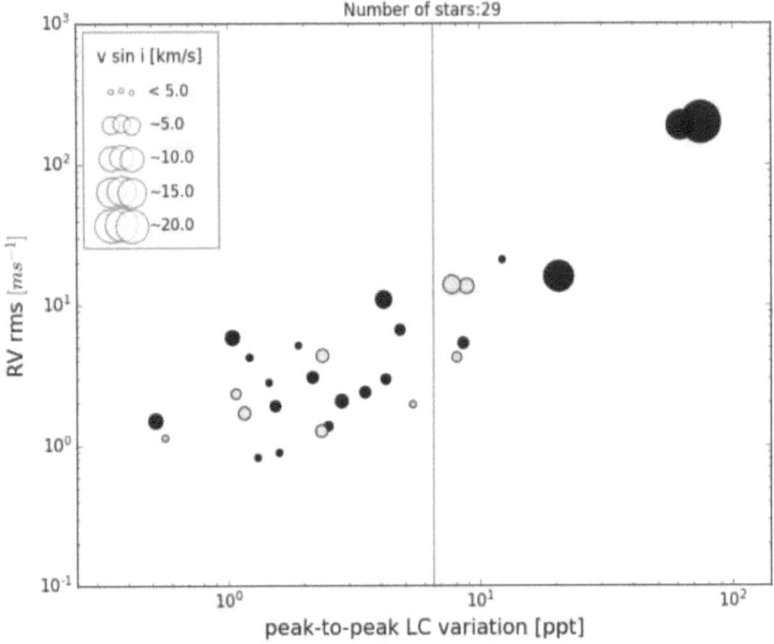

Figure 1.21: RV-RMS and the peak-to-peak of light curve variation for the subsample of 20 stars with spot dominance and 9 stars with faculae dominance in their light curves. Circle sizes represents the *v sin i* value determined spectroscopically. Black color indicates spot dominated and yellow indicates faculae dominated. The Blue vertical line shows the knee point peak-to-peak light curve variation at 6.5 ppt.

Main publications for the GPS method book:
The following chapters includes the three main manuscripts describing, proposing and testing the GPS method. Being this the core of my book disputation.

2 Inflection point in the power spectrum of stellar brightness variations: I. The model

2.1 Introduction of chapter 2

The magnetic features on stellar surfaces lead to quasi-periodic variations in stellar brightness as stars rotate. While such rotation variations were first detected with ground-based instrumentation (see, e.g., Radick et al. 1998), most of the data have been accumulated with planet-hunting spaceborne missions aimed at detecting planetary transits via photometric monitoring. In particular, CoRoT (Bordé et al. 2003b; Baglin et al. 2006) and *Kepler* (Borucki et al. 2010) telescopes provided photometric time series for several hundred thousand stars. Even more data is expected from the recently launched TESS mission (Ricker et al. 2014) and the future PLATO mission (Rauer et al. 2014).

The interest in studying stellar brightness variations is twofold. First, they provide information on the stars themselves, e.g. their rotation periods or their magnetic cycles. Second, a quantitative assessment of stellar variability is needed for better detection and characterization of extra-solar planets.

Of particular interest are studies of stellar rotation periods. Stellar rotation is closely linked to stellar magnetic activity and age (Skumanich 1972). Consequently, surveys of stellar rotation periods are the basis for calibrating gyrochronology relationships between rotation period, color, and age (cf. McQuillan et al. 2014), for understanding the Galactic star formation history (cf. Davenport 2017; Davenport and Covey 2018), and for constraining properties of the magnetic braking (Metcalfe et al. 2016). The light curves of many, especially young and active stars, look almost like a sine wave (see, e.g. Fig. 4 from Reinhold et al. 2013). The rotation period of such stars manifests itself as a clear peak in the Lomb-Scargle periodogram (Zechmeister and Kürster 2009) or a series of equidistant peaks in the autocorrelation function (McQuillan et al. 2013) of their light curves. Consequently, in the numerous studies aimed at determining stellar rotation periods employing *Kepler* and CoRoT data (Walkowicz and Basri 2013; Reinhold et al. 2013; McQuillan et al. 2014; García et al. 2014; Buzasi et al. 2016a; Angus et al. 2018) most of the obtained rotation periods are of such stars.

The largest available surveys of stellar rotation periods have been compiled by Reinhold et al. (2013) using Lomb-Scargle periodograms and by McQuillan et al. (2014) using autocorrelation analysis. They determined rotation periods in, respectively, 24124 and 34030 presumably main sequence *Kepler* stars. Another approach was taken by García et al. (2014), who concentrated on *Kepler* stars with measured pulsations and, in addition, to the autocorrelation analysis employed wavelet power spectra. They determined rotation periods in 310 out of 540 considered targets.

As successful as they are, the aforementioned approaches are based on the assumption that stellar light curves have a regular temporal profile. This is a valid assumption for young and active stars but it fails for many old and less active stars. For such stars the complex configuration of magnetic features and their relatively rapid evolution lead to rather complex light curves and render the period determination very difficult.

The most prominent example of such a star with complex light curve is our Sun. Solar short-term variability (i.e. variability on timescales of up to a few solar rotation periods) has a highly irregular temporal profile (see, e.g. Fig. 1 from Shapiro et al. 2016). The main reason for this is that only very few sunspots last longer than the solar rotation period so that the short-term variability of solar brightness is strongly affected by sunspot evolution. Furthermore, Shapiro et al. (2017) showed that the global wavelet power spectrum of solar brightness variations, calculated over the period 1996-2015, does not have a clear rotation peak due to the compensation of facular and spot contributions to solar brightness variability. They showed that since faculae have much longer lifetimes than spots (e.g. facular features can easily last for a few solar rotations), their contribution to solar brightness variability has a very pronounced peak at the solar rotation period.

The peak in the spot component of solar brightness variations is much less pronounced but the spot component is stronger on timescales around the solar rotation period (i.e. at about 10–50 days). As a result, two peaks almost fully cancel each other. This is in agreement with other studies (see, e.g. Lanza and Shkolnik 2014; Aigrain et al. 2015) that found that the true rotation period of the Sun would not be detectable at intermediate and high levels of solar activity, when the spot contribution to solar brightness variations wipes out the rotation peak in the facular component. At the same time, the solar rotation period is easily detectable during activity minimum when the brightness variations are brought about by long-lived faculae.

A good understanding of the physical phenomena determining solar variability might be helpful for solving problems posed by stellar data. For example, Reinhold et al. (2019) suggested that the dearth of the intermediate stellar rotation periods in the *Kepler* sample (see, e.g. McQuillan et al. 2014; Davenport 2017; Davenport and Covey 2018) can be partially caused by the compensation of facular and spot contributions to brightness variability (similar to the solar case) and the consequent inability to detect rotation periods of such stars. Therefore the dearth in the *observed* period distribution does not necessarily implies an under-representation in the real period distribution.

We suggest that the irregularity of stellar light curves is an important factor in explaining why rotation periods cannot be determined for the majority of stars in the *Kepler* field (e.g. the success rate of McQuillan et al. (2014) is only 25.5% since they applied the autocorrelation method to 133030 stars). Recently, van Saders et al. (2019) showed that the success rate of period determinations strongly decreases with increasing stellar effective temperature and such a decrease cannot be explained by the simultaneous decrease of the amplitude of stellar brightness variations. This is in line with our suggestion, since spot lifetimes are expected to decrease with stellar effective temperature (Giles et al. 2017) and, consequently, the period determination gets more difficult.

In this paper we employ an approach similar to that taken by the SATIRE model (which stands for Spectral And Total Irradiance Reconstruction, Fligge et al. 2000a; Krivova et al. 2003), originally developed for modeling solar brightness variations, to synthesize stellar light curves and their power spectra. We do this as a function of lifetimes of spots and faculae, the ratio between facular and spot stellar surface-area coverage, and stellar inclination (i.e. the angle between the direction to the observer and stellar rotation axis). We specify the conditions under which the rotation peak in stellar power spectra disappears and show that even in such cases the rotation period can still be determined from the high-frequency tail of the power spectrum. In particular, we calculated the frequency where the concavity of the power spectrum plotted on the log-log scale changes sign (in other words the steepest point of the power spectrum). Such a frequency corresponds to the inflection point in the power spectrum of stellar brightness variations. We demonstrate that the position of the inflection point is proportional to the stellar rotation frequency and can be used as a proxy for its determination. All in all, we show that the power spectrum of stellar brightness variations is a sensitive tool for studying stellar rotation and magnetic activity.

The SATIRE approach has been extensively validated against various solar data (see e.g. Ball et al. 2014; Yeo et al. 2014, and references therein). Recently SATIRE was used to show that observed solar brightness variations can be explained with remarkable accuracy by the joint action of only two sources, the surface magnetic field and granular convection (Shapiro et al. 2017). This result puts us in a strong position for modeling brightness variations of Sun-like stars.

This paper is restricted to modeling for validating the proposed approach, while its application to available stellar data is the subject of forthcoming papers. The rest of the paper is structured as follows: in Sect. 2.2 we describe the model used to synthesize the light curves presented in this study. In Sect. 2.3 we consider an illustrative case of stars whose variability is exclusively brought about by dark spots. A more realistic case of stars with dark spots and bright faculae is detailed in Sect. 2.4. The impact of various properties of stellar magnetic features on the position of the inflection point is outlined in Sect. 2.5, while the dependence of the inflection point position on the level of stellar magnetic activity is presented in Sect. 2.6. Finally, conclusions are drawn in Sect. 2.7.

2.2   Model description

Strong concentrations of magnetic field emerging on the stellar surface lead to the formation of active regions, encompassing magnetic features such as dark spots and bright faculae (see, e.g., review by Solanki et al. 2006). The transits of these regions over the visible stellar disk as the star rotates as well as their evolution are dominant sources of brightness variations in Sun-like stars on timescales from about a day.

In our model we construct active regions as a mixture of spot and facular areas. We note that a model based on such an assumption would not be suitable for calculating stellar brightness variations on the timescale of the magnetic activity cycle since it does not account for the emergence of the ephemeral active regions (see, e.g. Dasi-Espuig et al. 2016, and references therein). At the same time the variability on the timescale of stellar rotation is brought about by the largest facular and spot features, which usually emerge together. Therefore such a simple model of stellar active regions is expected to be appropriate for modeling stellar rotation variability and is often employed in the literature (see, e.g. Lanza et al. 2003, 2009; Gondoin 2008; Borgniet et al. 2015; Morris et al. 2018).

The size of active regions is assumed to be much smaller than the stellar radius, which is a good assumption for the Sun and for stars with near-solar levels of magnetic activity. Consequently, we did not consider the exact geometrical shape of active region and its spot and facular components, prescribing the same value of the foreshortening factor for the entire region when computing its visible solid angle.

At each moment of time t the stellar brightness at the wavelength λ is given by

$$F(\lambda, t) = F_Q(\lambda) + \sum_i \Omega_i(t) \cdot (\phi_{i,F}(t) C_F(\lambda, \vec{r_i}) + \phi_{i,S}(t) C_S(\lambda, \vec{r_i})), \qquad (2.1)$$

where $F_Q(\lambda)$ is the brightness of a quiet star, i.e. a star without any active regions. The summation is performed over all active regions visible at time t and $\Omega_i(t)$ is a solid angle of the i-th active region seen from the vantage point of observer. By changing the number of regions on the stellar surface we could simulate stars with different magnetic activity. Factors $\phi_{i,F}(t)$ and $\phi_{i,S}(t)$ are fractions of spot and facular parts of the area of i-th active region, respectively (with $\phi_{i,F}(t) + \phi_{i,S}(t) = 1$). $C_F(\lambda, \vec{r}_i)$ and $C_S(\lambda, \vec{r}_i)$ are spectral contrasts of faculae and spots relative to the quiet stellar regions along the direction to the i-th active region $\vec{r}_i$.

For simplicity, we consider here only stars rotating as a solid body. We assume that the emergence of stellar active regions follows the solar latitudinal distribution and happens in the activity belts, i.e. in between the latitudes of 5° and 30° (see, e.g. Knaack et al. 2001). This should be a good approximation for slowly rotating stars like the Sun (Schüssler and Solanki 1992), which are the main focus of our study.

As will be shown below the growth phase of the active regions does not have a strong effect on our results so that it is neglected in most of the experiments. In other words, we start tracking the region and its effect on stellar brightness only after it reaches its maximum area.

We have utilized spectra of the quiet Sun, faculae, spot umbra, and spot penumbra calculated by Unruh et al. (1999) with the ATLAS9 radiative transfer code (Kurucz 1992; Castelli and Kurucz 1994). Following Wenzler et al. (2006) and Ball et al. (2012) we compute sunspot spectra as a mixture of 80% penumbral and 20% umbral spectra.

The Unruh et al. (1999) spectra have been proved to be reliable for modeling solar brightness variations (see e.g. reviews by Ermolli et al. 2013; Solanki et al. 2013, and references therein) and, consequently, we expect them to be applicable to modeling stars with near-solar fundamental parameters. At the same time, the profile of the high-frequency tail of the power spectrum and, consequently, the position of the inflection point depends on the the centre-to-limb variations of brightness contrasts of stellar magnetic features. They, in turn, depend on the fundamental stellar parameters. The apparatus for calculating contrasts of magnetic features in stars with various fundamental parameters is becoming available (see Norris et al. 2017; Witzke et al. 2018; Salhab et al. 2018) so that we plan to generalize our study in one of the forthcoming publications. Presently available simulations of facular contrasts at stars with different effective temperatures (see, e.g., Fig. 5.16 from Norris et al. 2017) indicate that results presented in this study are applicable to at least stars from late F to early K spectral types.

All the light curves presented in this study are calculated as they would be seen by the *Kepler* telescope, i.e. by multiplying Eq. 2.1 with the *Kepler* total spectral efficiency and integrating it over all relevant wavelengths. The simulations are performed with 6-hour cadence. We have checked that the decrease of the time step in our simulations has virtually no effect on the spectral power of variability at periods from about 2–3 days and larger so that such a choice is appropriate for our goals.

As illustrated by Eq. 2.1 the variability of the flux $F(\lambda, t)$ is brought about by the time-dependence of the solid angles of active regions seen from the vantage point of observer, Ω_i, and by the time-dependence of facular and spot fractions $\alpha_{i,F}(t)$ and $\alpha_{i,S}(t)$. The former is attributed to the evolution of magnetic features as well as to the rotation of the star and consequent change of the foreshortening factor. The latter is given by the difference between facular and spot lifetimes. For example, in the (rather unrealistic) case of the same lifetime of spot and facular components of an active region, their relative coverage would not depend on time and consequently the time dependence of the contribution of active regions to stellar brightness will be solely determined by the variable solid angle $\Omega_i(t)$.

All in all, the power spectra of stellar brightness variations depend on properties of stellar active regions and the viewing geometry. To better illustrate the important effects and individual roles of each of the involved parameters we start with considering a greatly simplified case of variability in Sect. 2.3 and add more realism into our simulations in the subsequent sections (while still keeping the model relatively simple).

2.3   Stars with spots

In this section we examine stellar brightness variations due to spots, i.e. we put $\alpha_S(t)$ to 1 and $\alpha_F(t)$ to 0 in Eq. 2.1. We track a star during a time interval of 1600 days which roughly corresponds to the duration of 17 *Kepler* quarters, i.e. approximately the total duration of the *Kepler* mission. During this interval we let 300 spots emerge, each at a random point of time and in a random place within the activity belts on the stellar surface (see Sect. 2.2). Since we are interested in the impact of spot evolution on the profile of the power spectrum of stellar brightness variations and, in particular, on the position of the inflection point, for illustrative purposes we assume that all spots have the same growth and lifetimes, independently of their size. We also start with a simple case of spots emerging in three relative sizes scaling as 1:2:3 and consider 100 spots of each of the sizes. A more realistic treatment will be employed in the subsequent sections. We note that the absolute size of spots does not play a role in the calculations presented in this section since it affects only the amplitude of the brightness variations and has no effect on the profile of their power spectrum.

2.3.1   High-frequency tail of the power spectrum and inflection point

Figures 2.1 and 2.2 show two realizations of light curves calculated for a model star rotating with a 30-day period. We assumed that spots instantaneously emerge on the stellar surface (i.e. that the growth time is zero) and then their areas linearly decrease with time. In other words, the spot area $A(t)$ after the emergence can be written as

$$A(t) = A_0 \left(1 - \frac{t - t_0}{T_{\text{spot}}} \right), \quad t_0 \le t \le t_0 + T_{\text{spot}}, \tag{2.2}$$

where A_0 is the maximum area and t_0 is the time of emergence. We put $T_{\text{spot}} = 25$ d to produce light curves for Figs. 2.1 and 2.2. Since times and positions of individual emergence are kept random, the two light curves shown in these figures are distinctly different from each other.

One can clearly see the individual dips caused by the transits of spots as a star rotates (Figs 2.1a and 2.2a). Nevertheless, the Lomb-Scargle periodograms of both light curves do not have a clear 30-day peak (Figs 2.1b and 2.2b). Instead, the peaks appear to be rather random and their locations depend on the specific realization of spot emergence. The same situation is seen when global wavelet power spectra with 6th order Morlet and Paul wavelets (see Figs 2.1c and 2.2c and Figs 2.1e and 2.2e, respectively) are computed: all four power spectra do not have any noticeable signature of the rotation peak. In comparison to the Morlet wavelet, Paul wavelet implies a poorer frequency localization but stronger averaging in the frequency domain when power spectra are computed. Consequently, wavelet power spectra calculated with the Paul wavelet have less details but they are more resistant to statistical noise (Torrence and Compo 1998).

In Figs 2.1d, f and 2.2d, f we plot the ratios R_k between the power spectral density $P(\nu)$ at two adjacent frequency grid points: $R_k \equiv P(\nu_{k+1})/P(\nu_k)$. It is easy to show that these ratios can be written as

$$R_k = 1 + \frac{d \ln P(\nu_k)}{d \ln \nu} \cdot \frac{(\Delta \nu)_k}{\nu_k}, \tag{2.3}$$

where $\Delta \nu$ is spacing of the frequency grid. We calculate power spectra on a grid that is equidistant on a logarithmic scale, i.e. $\Delta \nu / \nu$ is constant. Therefore, R_k values represent gradient of the power spectrum plotted on a log-log scale (as in Figs 2.1c, e and 2.2c, e), scaled with some factor (which depends on the chosen frequency grid) and offset by unity.

For simplicity from now on we will refer to the R_k values as the gradient of the power spectrum. One can see that while the gradient of the Morlet power spectra has sophisticated profiles with many local maxima (corresponding to inflection points in the Morlet power spectrum), the gradient of the Paul power spectra looks much simpler. Furthermore, while the power spectra of both light curves have no noticeable peak at the rotation period, both 6th order Paul power spectra have inflection points giving rise to very clear peaks in the gradients of the power spectra. Importantly, the location of these points is the same for the two realizations plotted in Figs. 2.1 and 2.2.

In Figs. 2.1 and 2.2 we show two light curves corresponding to the same lifetime of spots, but to different realizations of spot emergence. In Fig. 2.3 we look at things the other way around and consider power spectra of four light curves calculated with the same realization of spot emergence but with different lifetimes of spots. The power spectrum of the light curve with a spot lifetime $T_{\rm spot} = 90$ days has a pronounced rotation peak. Its amplitude decreases rapidly with decreasing spot lifetime and disappears completely when the lifetime of spots becomes smaller than the stellar rotation period: neither $T_{\rm spot} = 20$ days nor $T_{\rm spot} = 12$ days cases display any signature of the peak in the power spectrum around the rotation period. To better illustrate this point we also plot the power spectra on a linear vertical scale (Fig. 2.3b).

Figure 2.3 illustrates that stellar rotation periods cannot be determined from the maximum of the power spectrum when lifetimes of spots are small in comparison to the rotation period (at least for a star with no faculae). Interestingly, this is the case for the Sun since sunspots very rarely last longer than the solar rotation period (Baumann and Solanki 2005).

The bottom panel of Fig. 2.3 points to an alternative method for determining rotation periods when spot lifetimes are shorter than the stellar rotation period. One can see that the high-frequency tail of the power spectra is much less sensitive to spot lifetime. In particular, the position of the inflection point is almost the same in all four cases. The results obtained so far strongly suggest that the high-frequency tail of the power spectrum may provide a more robust way of determining stellar rotation periods. There are different ways of parameterizing the tail, e.g. one can approximate it with the help of a multi-component powerlaw fit similar to that employed by Aigrain et al. (2004) and establish the connection between parameters of such a fit and the rotation period. However, in the present study we limit ourselves to showing that the position of the inflection point is a sensitive proxy of the stellar rotation period, leaving other methods for future investigations.

The profile of the power spectrum and, consequently, the calibration factor between the position of the inflection point and rotation period depend on the wavelet utilized for

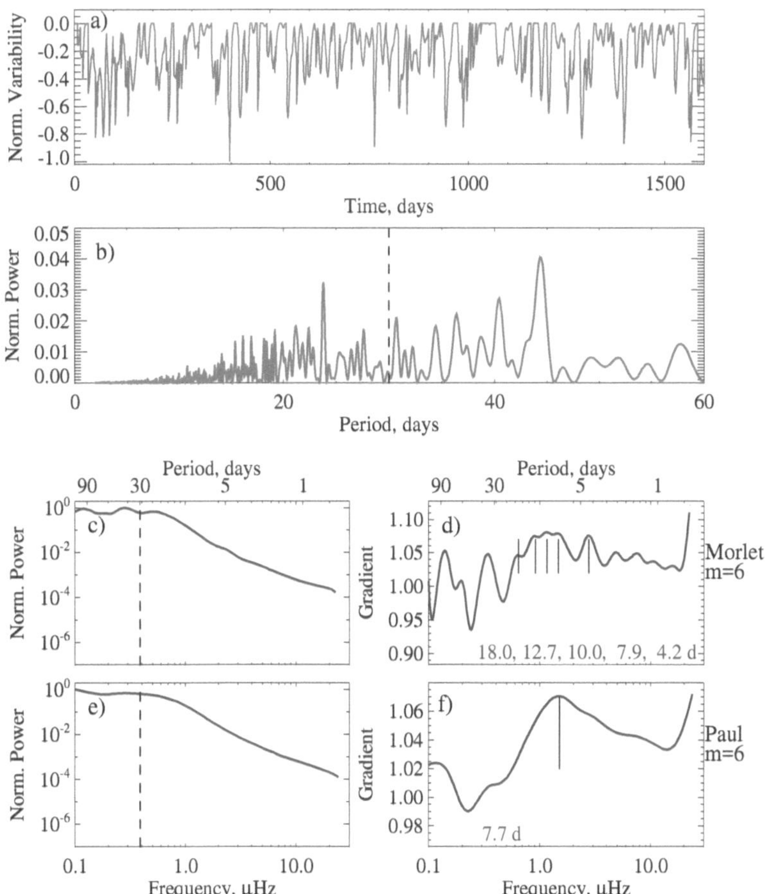

Figure 2.1: Model light curve of a star with a 30-day rotation period covered by spots and observed from its equatorial plane. The spots decay according to a *linear* law with $T_{\mathrm{spot}} = 25$ d. Two upper panels show a normalized light curve (panel a) and corresponding Lomb-Scargle periodogram (panel b). Panels c–f show global wavelet power spectra (left panels) and corresponding gradient of the power spectra (right panels) calculated with the 6th order Morlet wavelet (panels c and d) and with the 6th order Paul wavelet (panels e and f). The values of the gradients of these power spectra are a scaled and offset by unity (see Eq. 2.3 and discussion in the text for the exact quantity plotted). Numbers in panels d and f correspond to the positions of the inflection points (i.e. local maxima of the gradient). Vertical dashed lines in panels b, c, and e indicate the rotation period of the modeled star. Vertical solid lines in panels d and f indicate positions of the inflection points. We note that since spots reduce stellar brightness, the normalized variability (i.e. normalized $F(\lambda, t) - F_{Q}(\lambda)$ values, see Eq. 2.1) is plotted between -1 and 0.

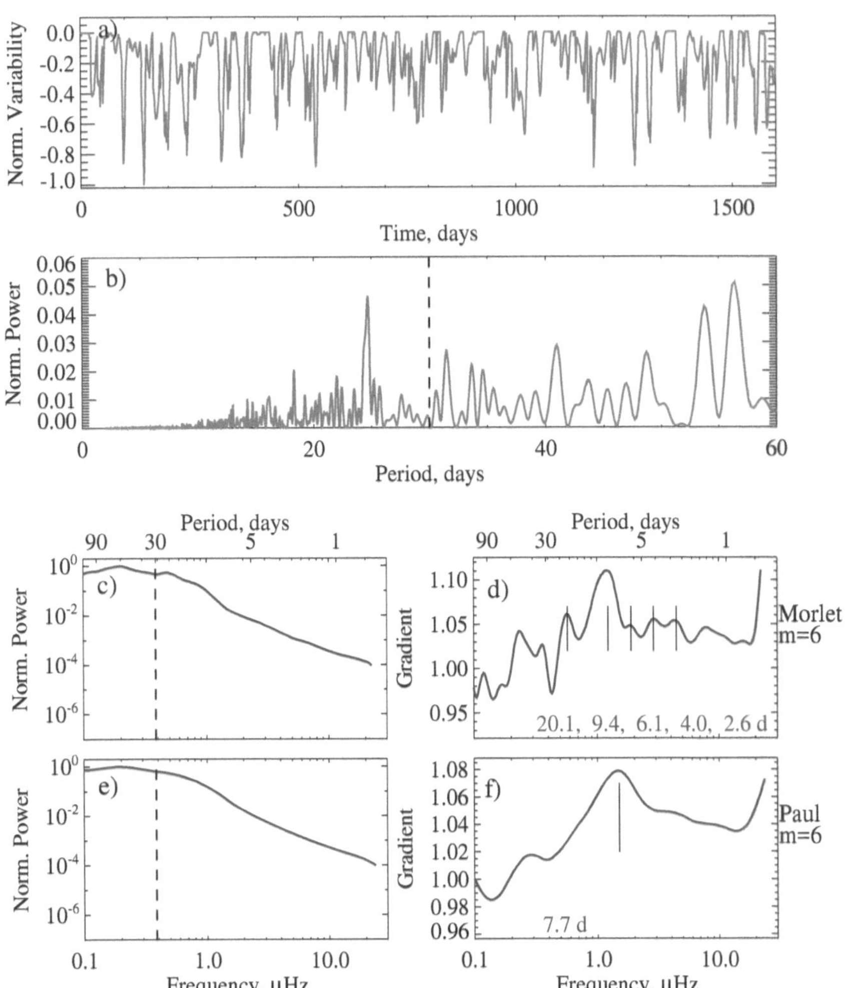

Figure 2.2: The same as Fig. 2.1 but for another realization of spot emergence.

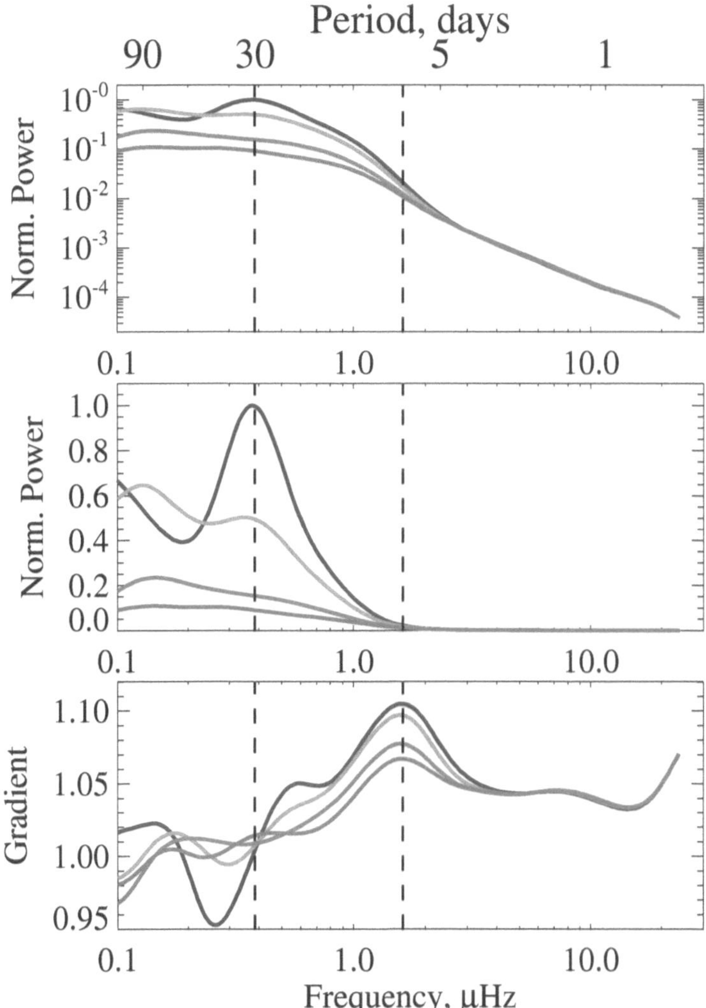

Figure 2.3: Power spectra of model light curves plotted on logarithmic (top panel) and linear (middle panel) scales on the vertical axis. The gradients of the power spectra in the top panel are plotted in the bottom panel. The modeled spots decay according to a *linear* law with lifetimes, T_{spot}, equal to: 90 d (blue), 50 d (orange), 20 d (magenta), and 12 d (red). All four light curves are calculated for the same realization of spot emergence. Vertical dashed lines at 30 d and 7.2 d correspond to the rotation period of the simulated star and the approximate position of the inflection point in all four power spectra, respectively. Power spectra are calculated with the 6th order Paul wavelet.

calculations. The wavelets with good frequency localization lead to power spectra with multiple, often many inflection points whose positions depend on the specific realization of emergence (compare Figs. 2.1d and 2.2d). At the same time wavelets with very low frequency localization lead to a strong scatter in the relationship between inflection point position and the rotation period. After considering several wavelets with different degrees of frequency localization we found that the 6th order Paul wavelet introduces the best smoothing of the power spectra for our purposes. An example of power spectra and corresponding gradients calculated utilizing wavelets with different frequency localization is given in Fig. 2.12.

We stress that the inflection point itself does not have a clear physical meaning and it is just a convenient way of quantifying the profile of the high-frequency tail of the power spectrum.

2.3.2   Effect of spot emergence and lifetime

In Fig. 2.4 we show the dependence of the inflection point position on the lifetime of spots. In addition to the linear decay law, several other functional forms, such as parabolic and exponential decays (Bumba 1963; Martínez Pillet et al. 1993; Petrovay and van Driel-Gesztelyi 1997) have been proposed (see also Solanki 2003, for a review). To illustrate the impact of the functional form of the decay law on the inflection point position we also consider an exponential law under which the spot area can be written as

$$A(t) = A_0 \exp(-\frac{t - t_0}{T_{\text{spot}}}), \quad t \geq t_0. \tag{2.4}$$

Figure 2.4 shows that the position of the inflection point remains stable and is not affected by T_{spot} for values above about 15 d (i.e. one half of the rotation period) for the linear decay law and 10 d (one third of the rotation period) for the exponential decay law. Note, however, the slightly different meanings of T_{spot} for the two decay laws. T_{spot} corresponds to spot lifetime for linear decay (i.e. the spot disappears when T_{spot} is reached), while T_{spot} in the exponential decay law implies an e-folding time.

Since the exact profile of the power spectrum depends on the realization of spot emergence, the positions of the inflection point show some scatter around the mean values (solid lines in Fig. 2.4) for a fixed lifetime. This scatter represents an intrinsic uncertainty of using the inflection point as a proxy for the rotation period. For some of the realizations "rogue" inflection points at lower periods appear (seen below 0.1 in the upper panel). Furthermore, inflection points are also found at large periods, even forming a high-period branch for spot lifetimes larger than about 30 days. These points are linked to the rotation peak in the power spectra (which is only present if spot lifetime is large, see blue curves in Fig. 2.3).

Until now we considered spots that emerge instantaneously on the stellar surface and then decay. Such an assumption is reasonable for our purposes since the emergence and growth of spots takes significantly less time than the decay and rarely lasts longer than a few days (see, e.g. van Driel-Gesztelyi and Green 2015). To estimate the effect of the non-zero growth time we compare in Fig. 2.5 the positions of the inflection points calculated for a spot emergence (and growth) time of 0 d (i.e. assuming instantaneous emergence as in Fig. 2.4), 1 d, and 2 d. We assumed a linear growth of spot area during the

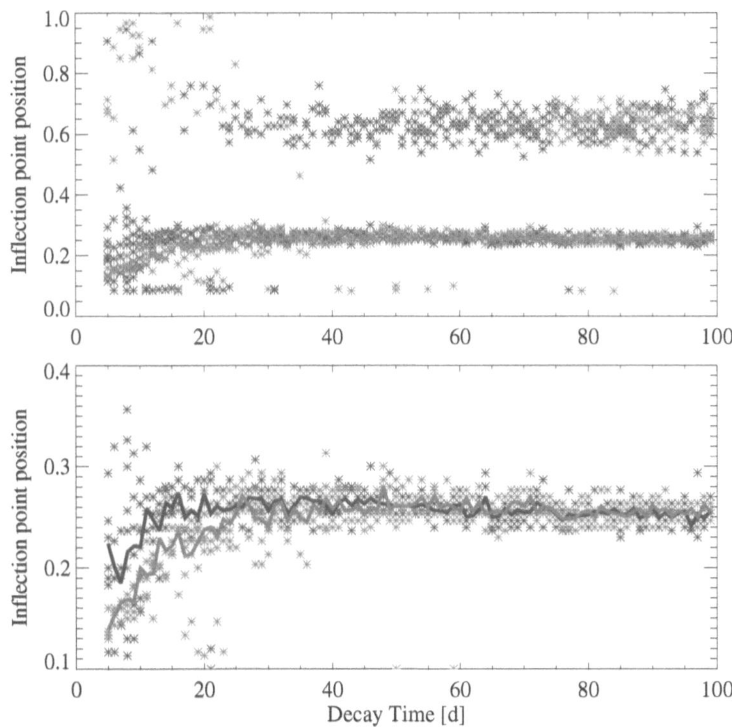

Figure 2.4: Dependence of the inflection point position (in fractions of the rotation period, P_{rot} = 30 d) on spot lifetime, T_{spot}. Each value of the spot lifetime corresponds to five realizations of spot emergence with linear (red asterisks) and five realizations with exponential (blue asterisks) decay laws. Lower panel is a zoom in of the upper panel. Blue and red lines in the lower panel show positions of the high-frequency inflection points averaged over corresponding five realizations.

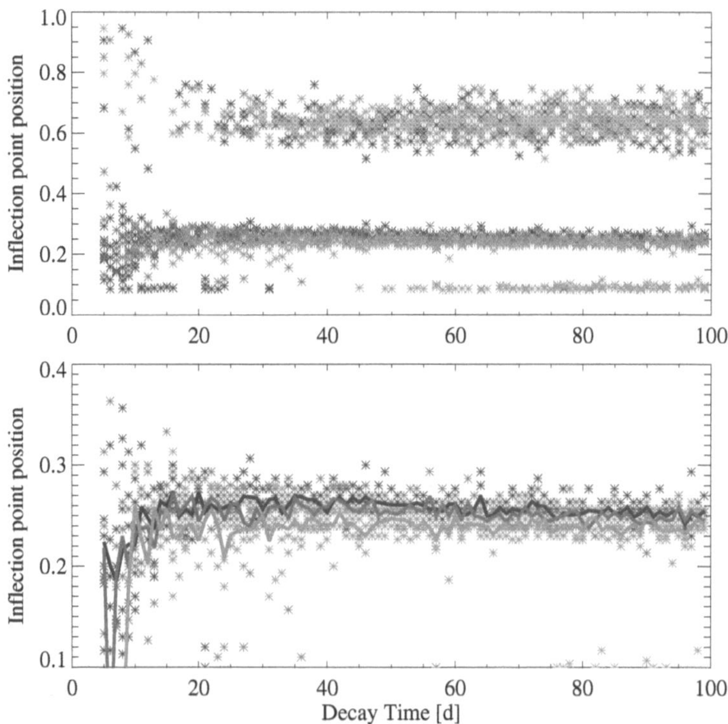

Figure 2.5: The same as Fig. 2.4, but now comparing inflection points obtained for three values of spot emergence time: 0 d (blue), 1 d (red), and 2 d (orange). An exponential decay law is imposed.

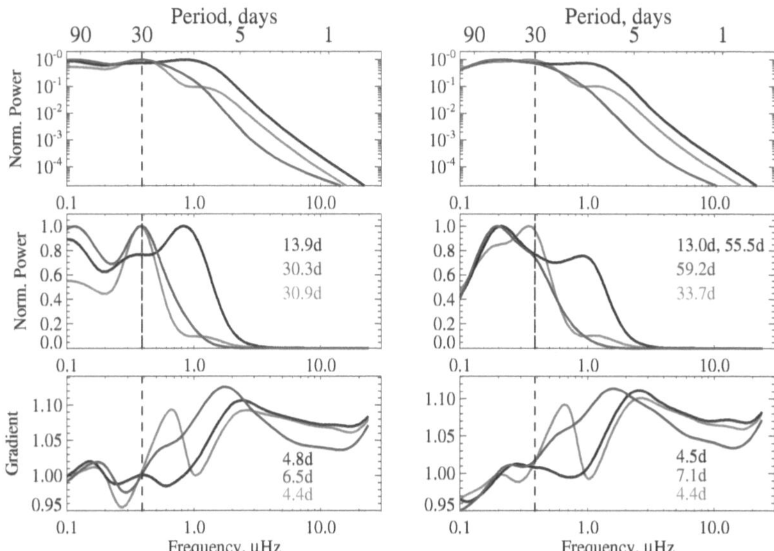

Figure 2.6: Power spectra of brightness variations of a star rotating with a 30-day period plotted on a logarithmic (upper panels) and linear (middle panels) scales. Also plotted are the gradients of the power spectra (lower panels). The calculations presented in left and right hand panels have been performed with the same set of model parameters (see text for details) but are associated with two different realizations of active regions emergence. Black, red, and blue curves correspond to total brightness variations (calculated with $S_{\text{fac}}/S_{\text{spot}} = 5.5$), as well as their facular and spot components, respectively. Numbers in the middle panels indicate peaks in the power spectra, while numbers in the lower panels point to positions of the inflection points.

emergence phase. One can see that including a non-zero emergence phase slightly shifts the inflection point to lower periods, but the effect is relatively small (on average 8% for 2d emergence time).

2.4   Stars with spots and faculae

Sunspots are generally parts of bipolar magnetic regions, which also harbor smaller magnetic elements. Ensembles of these magnetic elements form bright faculae (see, e.g. Solanki et al. 2006; Solanki et al. 2013, for reviews). Faculae are present on late-type stars and play an important role in stellar photometric variability (see, e.g. discussion in Shapiro et al. 2016; Witzke et al. 2018; Reinhold et al. 2019). For example, faculae dominate the variability over the course of magnetic activity cycles for old stars, like the Sun (Lockwood et al. 2007; Radick et al. 2018). They also significantly affect solar brightness variations on timescales of a few days (Shapiro et al. 2016, 2017) and thus one can expect that the position of the inflection point in power spectra of stars similar to the Sun is affected by the facular contribution to stellar brightness variability. In this section we investigate the effect of faculae on the position of the inflection point in the power spectrum of stellar brightness variations.

2.4.1   Treatment of faculae

Here we extend the model outlined in Sect. 2.2 and describe the treatment of the facular contribution to stellar brightness variability. Furthermore, we relax the assumption of the equal lifetime for all magnetic features adopted in Sect. 2.3 for illustrative purposes. Instead we consider a more comprehensive model of the decay of magnetic features.

For simplicity we limit ourselves to the case of the instantaneous emergence of active regions. This should not affect any of the conclusions drawn here since the duration of the emergence does not have a strong impact on the position of the inflection point (see Sect. 2.3.2). We assume that immediately after the emergence all magnetic regions have the same fractional coverage by spot and facular components. Consequently, we calculate the power spectrum of photometric variations and position of the inflection point as a function of the facular to spot area ratio *at the time of maximum area*, S_{fac}/S_{spot}. We note that for the case of the instantaneous emergence the time of maximum area coincides with the time of emergence. Since facular and spot lifetimes are generally different the ratio *at the time of maximum area*, S_{fac}/S_{spot}, is not identical to the *instantaneous* (i.e. *snapshot*) ratio obtained at any random instance.

We adopt a solar log-normal distribution of spot sizes, taken from Baumann and Solanki (2005), but only considering spots larger than 60 MSH (micro solar hemisphere). Consequently, the size of the spot component of each emerging magnetic region was randomly chosen following the Baumann and Solanki (2005) distribution. The log-normal distribution implies that while most of the spots have small sizes of about 100 MSH, every now and then huge spots with sizes of more than 3000 MSH appear. Then instead of considering a constant lifetime of all spots as we did in Sect. 2.3 we follow Martínez Pillet et al. (1993) and consider a constant decay rate of spots. This results in a linear decay law with large spots living longer than small spots. The choice of the decay rate is not straightforward since it is rather poorly constrained even in the solar case. We will consider values between 10 MSH/day given by Gnevyshev-Waldmeier relation between sunspot sizes and lifetimes (Waldmeier 1955) and 41 MSH/day given by Martínez Pillet et al. (1993). In any case, as will be shown below, the position of the inflection point is basically independent of the decay rate.

The lifetimes of spots are computed from spot areas and decay rates. To calculate the lifetime of faculae we assume a fixed ratio between lifetimes of facular and spot components of the active region (which implies a fix decay rate also for faculae). Since lifetimes of the facular component are usually significantly larger than those of spots (see, e.g. reviews by Solanki et al. 2006; van Driel-Gesztelyi and Green 2015) the active regions in our model emerge as a mixture of spot and facular regions, but then spend a significant part of their lifetimes as purely facular regions.

In our simplified parametric consideration of active region evolution we do not directly account for the faculae brought about by the decay of spots. Faculae from sunspot decay imply a) underestimation of facular areas in our model; b) deviations of the facular decay law from linear. Point a) can be indirectly taken into account by the increase of the $S_{\text{fac}}/S_{\text{spot}}$ ratio (in other words facular area in this ratio represents not only the facular features emerging together with spots but also the product of the spot decay). We also do not expect that point b) can noticeably affect our calculations since the exact time evolution of magnetic features does not have a strong impact on the position of the inflection point. Recently Işık et al. (2018) performed more realistic calculations of magnetic flux emergence and surface transport in stars with various rotation periods. As a next step we plan to employ their results in our modelling.

2.4.2   Superposition of spot and facular contributions to stellar brightness variability

In Fig. 2.6 we depict power spectra of brightness variations brought about by faculae, by spots, and by their mixture (red, blue, and black lines, respectively). We have put spot decay rate to 25 MSH/day, i.e. roughly in between the estimates given by Waldmeier (1955) and Martínez Pillet et al. (1993). The facular components of active regions were set to live twice as long as spot components. We have considered 1600-day light curves and let 2400 emergence randomly happen during this time. The absence of any clustering of emergence in time implies that the mean activity level of a star during the entire time of simulations stays the same, i.e. we do not consider activity cycles. 2400 emergence resulted in a mean fractional disk spot coverage being about 0.3% (due to the adopted log-normal distribution of spot sizes the exact value slightly depend on the specific realization of emergence), which is a typical solar value around the activity maxima.

Left and right panels of Fig. 2.6 show power spectra of two light curves as well as of their facular and spot components. Both light curves have been calculated with the same set of model parameters specified above, but correspond to two different realizations of magnetic region emergence. In the realization plotted in the left panels both spot and facular components have a prominent peak at the stellar rotation period. However, since facular and spot components are in anti-phase at periods around the rotation period (see discussion in Shapiro et al. 2017) the superposition of them leads to a disappearance of the rotation harmonic in the power spectrum of total brightness variations. Instead, a pronounced maximum in the power spectrum appears at 13.9 d, i.e. it is shifted by about 54% from the rotation period. The bottom panels of Fig. 2.6 shows that the location of the inflection point is different for the facular and spot components. This is not surprising since high-frequency tail of the power spectrum depends on the centre-to-limb variations of magnetic features contrasts and those are different for spots and faculae. In the given example the position of the inflection point of total brightness variations is shifted by 26% relative to the position of the inflection point of the spot component alone. This number corresponds to the error in determining the rotation period which will be made if, in the absence of any information about the relative role of spot and facular components of the variability, one connects rotation period and position of the inflection point assuming purely spot-dominated variability. We note that in the case presented in the left panel of Fig. 2.6 such an error is more than two times smaller than that made when assuming that rotation period corresponds to the maximum of the power spectrum (26% vs. 54%).

In the realization plotted in the right panels of Fig. 2.6 the spot component does not have a maximum at the rotation period, while the facular component still shows a clear

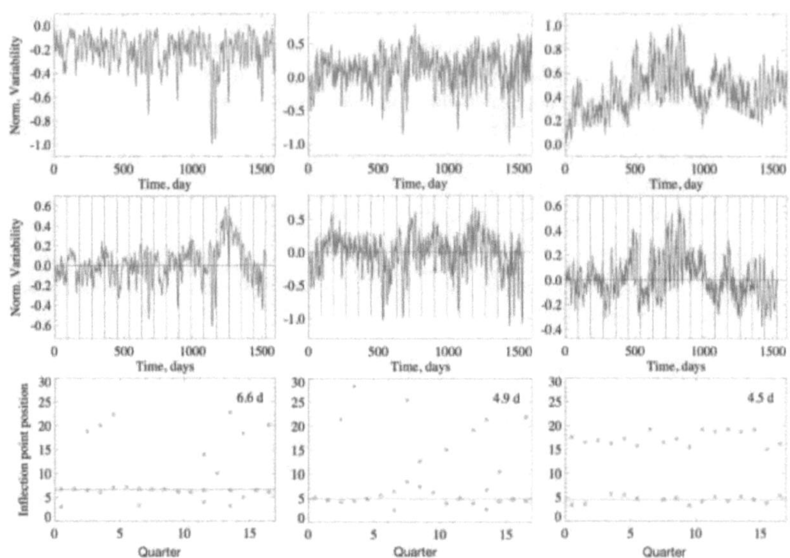

Figure 2.7: Three examples of simulated stellar variability: spot-dominated variability ($S_{\mathrm{fac}}/S_{\mathrm{spot}} = 0.01$, left panels), intermediate case ($S_{\mathrm{fac}}/S_{\mathrm{spot}} = 3$, middle panels), and faculae-dominated variability ($S_{\mathrm{fac}}/S_{\mathrm{spot}} = 100$, right panels). Upper panels show original (i.e. without any detrending) light curves. Intermediate panels show light curves split in 17 90-day quarters and linearly detrended in each of the quarters. The separation between quarters is marked by the vertical black lines. The asterisks in the lower panels correspond to the positions of inflection points in each of the quarters. Numbers in the upper right corners of the lower panels are the outlier-resistant mean values of the inflection point positions. These values are also indicated in the lower panels by horizontal black lines. Red asterisks correspond to the inflection points utilized for calculating the outlier-resistant mean value, blue asterisks are trimmed as outliers.

maximum (although slightly shifted to larger periods). In line with the discussion in Sect. 2.3 the position of the inflection point of the spot component is not affected by the disappearance of the peak corresponding to the rotation period (the shift of 0.6 d is within the scatter between different realizations of emergence, see Figs. 2.4–2.5). The superposition of the facular and spot components results in two peaks in the power spectra of total brightness variations, one at 13.0 d (i.e. shifted from the rotation period by 57%) and another at 55.5 d (i.e. shifted by 85%). Both numbers are larger than the shift of the inflection point caused by the facular component which is equal to 37%.

2.5 Main factors affecting position of the inflection point

In this section we investigate the dependence of the inflection point position on the facular to spot area ratio *at the time of maximum area*, S_{fac}/S_{spot}, (Sect. 2.5.1) and test this dependence against the solar case (Sect. 2.5.2). We also establish the dependence of the inflection point position on stellar inclination (Sect. 2.5.3).

In Sect. 2.3–2.4 we synthesized 1600-day light curves and then employed them for calculating power spectra and positions of the inflection points. While such a definition of the inflection point was appropriate for the illustrative purposes of Sect. 2.3–2.4, here we update the way the position of the inflection point is calculated to bring our calculations more into line with available stellar photometric data (e.g. *Kepler* or TESS)

As in Sect. 2.3–2.4 we synthesize 1600-day light curves but instead of employing them directly for calculating positions of the inflection points we first make "Kepler-like" light curves out of them. In other words, we split the light curves in 17 90-day quarters (ignoring the last 70 days) and linearly detrend each of the quarters. Then, instead of calculating the positions of the inflection points using the entire light curve, we calculate the positions of the inflection points in every quarter and consider the outlier-resistant mean, ignoring points outside of two standard deviations from the mean value.

This procedure is illustrated in Fig. 2.7 for spot- and faculae-dominated variability as well as for the intermediate case of the facular to spot area ratio *at the time of maximum area* (see Sect. 2.4.1), $S_{fac}/S_{spot} = 3$ (compare top and middle panels to see the difference between original and "Kepler-like" light curves). We have adopted a value of 25 MSH/day for the sunspot decay rate and set the facular lifetime to be three times that of spots. The positions of the inflection points in each of the quarters are plotted in the bottom panels of Fig. 2.7. The inflection points cluster in branches and, in particular, one can clearly see the branch corresponding to the high-frequency inflection point (i.e. at about 5–7 d). In the case of faculae-dominated variability, there is also a stable branch of low-frequency inflection points (in between 15 and 20 d, see right bottom panel of Fig. 2.7). This is due to the lifetime of faculae being sufficiently large for preserving a low-frequency inflection point (see discussion in Sect. 2.3.2). Since the high-frequency branch is more stable we constrain ourselves to its analysis and refrain from studying the low-frequency branch. The existence of the low-frequency branch might potentially lead to an ambiguity in the period determination. If, for example, the high-frequency branch is not visible due to the high noise level in the data then low-frequency branch might be erroneously taken for the high-frequency branch. This would lead to a roughly four times overestimation of the period. Such an ambiguity can be resolved by applying additional criteria. For example, one would expect that rotational periods of fast rotators should be caught by the autocorrelation or Lomb-Scargle periodograms techniques. Consequently, if there is, for example, an ambiguity between rotation periods of 7 and 28 days and both Lomb-Scargle periodograms and autocorrelation analysis fail, then the 28-day value should be chosen for the rotation period.

Figure 2.7 also shows that the positions of the inflection points slightly fluctuate from quarter to quarter and sporadically "rogue" inflection points appear. This is because the exact profile of the power spectrum depends on the specific realization of emergence of magnetic regions.

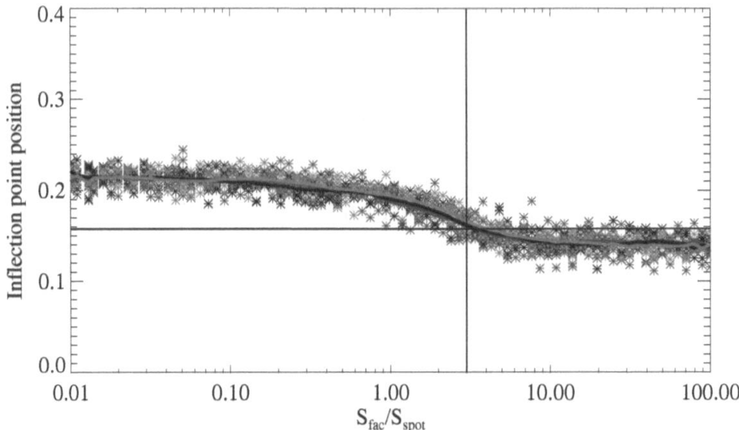

Figure 2.8: Dependence of the inflection point position (given as a fraction of the rotation period, P_{rot} = 30 d) on the facular to spot area ratio at the time of maximum area, $S_{\mathrm{fac}}/S_{\mathrm{spot}}$. Shown in red are calculations with lifetime of facular component of active regions, T_{fac}, equal to the lifetime of spot component, T_{spot}. Black and blue correspond to the $T_{\mathrm{fac}}/T_{\mathrm{spot}}$ = 2 and $T_{\mathrm{fac}}/T_{\mathrm{spot}}$ = 3 cases, respectively. For each pair of $S_{\mathrm{fac}}/S_{\mathrm{spot}}$ and $T_{\mathrm{fac}}/T_{\mathrm{spot}}$ values five realizations of emergence of active regions are shown. In other words, each of the $S_{\mathrm{fac}}/S_{\mathrm{spot}}$ values correspond to five red, five black, and five blue asterisks. Red, black, and blue lines mark the positions of the inflection points averaged over the five corresponding realizations. The black horizontal line indicates the position of the solar inflection point from Amazo-Gómez et al. (2020b), while the black vertical line marks the solar $S_{\mathrm{fac}}/S_{\mathrm{spot}}$ value established in Appendix 2.8.2.

2.5.1   Position of the inflection point as a function of the facular to spot area ratio

In Fig. 2.8 we present the dependence of the inflection point position on the area ratio between the facular and spot components of active regions *at the time of maximum area*, $S_{\mathrm{fac}}/S_{\mathrm{spot}}$. We keep the mean fractional disk-area spot coverage constant and set it to about 0.3% (see Sect. 2.4.2). Hence, the $S_{\mathrm{fac}}/S_{\mathrm{spot}}$ value affects only the facular coverage.

The decay rate of spots was chosen to be 10 MSH/day, i.e. according to the Gnevyshev-Waldmeier relation. In agreement with the calculations presented in Figs. 2.6–2.7 we have considered a fixed ratio between lifetimes of facular and spot components of active regions (T_{fac} and T_{spot}, respectively). We note that in the solar case the faculae last significantly longer than spots (see, e.g., review by Solanki et al. 2006). For example, Preminger et al. (2011); Dudok de Wit et al. (2018) found that facular features can affect solar UV irradiance (where it can be disentangled from noise more easily than in the white light) for up to 3–4 solar rotations (see their Fig. 5). In this context, Fig. 2.8 shows calculations for $T_{\mathrm{fac}}/T_{\mathrm{spot}}$ = 1, $T_{\mathrm{fac}}/T_{\mathrm{spot}}$ = 2, and $T_{\mathrm{fac}}/T_{\mathrm{spot}}$ = 3 cases.

Figure 2.8 shows that for spot-dominated variability (i.e. for small $S_{\mathrm{fac}}/S_{\mathrm{spot}}$ values) the inflection point is located at 22% of the rotation period (we only plot the high-frequency inflection points). We note that a small shift with respect to the 25% value seen in Figs. 2.4–2.5 is brought about by the different procedures for calculating the inflection point position, i.e. taking the outlier-resistant mean of 17 90-day intervals instead of computing a single inflection point. In the case of faculae-dominated variability (i.e. of large $S_{\mathrm{fac}}/S_{\mathrm{spot}}$ values) the inflection point is located at about 14% of the rotation period. The level of the statistical noise (i.e. variations in inflection point position caused by the random pattern of active regions emergence) is about 2–3%.

While the position of the inflection point strongly depends on the $S_{\mathrm{fac}}/S_{\mathrm{spot}}$ value, the difference between the three considered $T_{\mathrm{fac}}/T_{\mathrm{spot}}$ values is barely visible (compare red, blue, and black curves in Fig. 2.8). This has two important implications. First, auspiciously, the ambiguities in facular lifetime do not have a strong effect on the calculations of the inflection point position. Second, the position of the inflection point depends rather on the facular to spot area ratio *at the time of maximum area* than on the *instantaneous* ratio (which is proportional to the product of area ratio at the time of maximum area and ratio of the facular and spot lifetimes). We note that this result is in line with the discussion in Sect. 2.3, where we showed that the position of the inflection point only weakly depends on the lifetime of magnetic features.

Since stellar $S_{\mathrm{fac}}/S_{\mathrm{spot}}$ values are a priori unknown, their effect on the relation between rotation period and inflection point position introduces additional uncertainty in the period determined with the help of the inflection point (see Sect. 2.6 for a more detailed discussion). At the same time the dependence of the inflection point position on the $S_{\mathrm{fac}}/S_{\mathrm{spot}}$ value makes it possible to determine the ratio for stars with known rotation periods. We note that since the dependence presented in Fig. 2.8 is rather noisy, it is more suitable for studying general trends (e.g. the dependence of facular to spot ratio on stellar activity) than for deducing $S_{\mathrm{fac}}/S_{\mathrm{spot}}$ values for individual stars. We plan to determine $S_{\mathrm{fac}}/S_{\mathrm{spot}}$ values for McQuillan et al. (2014) sample of 34,030 stars with known rotation periods in a forthcoming publication. In this paper we limit ourselves to giving an example of application of the GPS method to stars significantly more variable than the Sun with presumably spot-dominated variability (see Appendix 2.8.3).

The calculations presented so far in this section have been performed for a fixed values of the rotation period, mean fractional disk-area spot coverage, and spot decay rates. In Appendix 2.8.1 we illustrate that the calibration factor between the inflection point position and rotation period is only marginally influenced by the rotation period (Fig. 2.13), spot coverage (Fig. 2.14), and spot decay rate (Fig. 2.15). Furthermore, we show that the position of the inflection point only barely depends on the latitude of the emerging active regions (Fig. 2.16).

2.5.2   Inflection point in the power spectrum of solar brightness variations

Let us now locate the Sun in Fig. 2.8. This requires knowledge of the inflection point position in the power spectrum of solar brightness variations as well as of the solar $S_{\mathrm{fac}}/S_{\mathrm{spot}}$ value. Amazo-Gómez et al. (2020b) demonstrated that the inflection point in the power spectrum of solar brightness variations is located at a period of about 4.17 days which is roughly 15.9% of the solar synodic rotation period at the equator. There have been also a number of studies aimed at determining the *instantaneous* ratio between facular and spot solar disk-area coverage (see, e.g. Chapman et al. 1997). At the same time the solar value of the facular to spot ratio *at the time of maximum area*, S_{fac}/S_{spot} is, on the whole, rather poorly studied and until now has remained unknown. In Appendix 2.8.2 we present a new method for determining the solar S_{fac}/S_{spot} value and show that mean solar value over the 2010–2014 period is about 3. Fig. 2.8 demonstrates that this value, in combination with the position of the solar inflection point from Amazo-Gómez et al. (2020b), agrees well with our calculations. This is reassuring, since it indicates that our simple parametric model allows accurate calculations of the inflection point position.

We remind that due to the lack of constraints on the dependence of S_{fac}/S_{spot} value on size of magnetic regions we assumed the same S_{fac}/S_{spot} ratio for all emerging magnetic regions. Solar data indicate that the *instantaneous* ratio between disk-area coverage by faculae and spot decreases from minimum to maximum of solar activity (Chapman et al. 1997; Foukal 1998; Solanki and Unruh 2013; Shapiro et al. 2014a). One can speculate that such a behavior is partly attributed to a stronger cancellation of small magnetic flux concentrations (associated with faculae) at higher levels of solar activity when regions with opposite polarities lie closer to each other (Cameron 2018, private communication). Based on this one can suggest that the ratio *at the time of maximum area* should not show as strong dependence on solar activity as the *instantaneous* ratio. This is in line with the results of Amazo-Gómez et al. (2020b), who could not pinpoint any clear dependence of the solar inflection point (which depends on the ratio *at the time of maximum area*, see above) on the level of solar activity. A possible changes of this ratio within a stellar activity cycle would contribute to the scatter in position of the inflection points.

2.5.3   Effect of inclination

The trajectories of active regions across the stellar disk as a star rotates depend on the position of the observer relative to the stellar equator. Consequently, stellar brightness variability is a function of the inclination (Schatten 1993; Knaack et al. 2001; Vieira et al. 2012; Shapiro et al. 2016), which is the angle between the stellar rotation axis and the direction to the observer. Therefore, one can expect that the position of the inflection point depends on the inclination.

Figure 2.9 is the same as Fig. 2.8, except the different colored symbols now represent different stellar inclinations. In contrast to Fig. 2.8, all calculations shown in Fig. 2.9 are performed with $T_{fac}/T_{spot} = 3$, but with three different values of the inclination: 90° (blue), 57° (black), and 45° (red). An inclination of 90° corresponds to observations from the equatorial plane (so that the blue asterisks are identical in Figs. 2.8 and 2.9). An inclination of 57° is the mean value of the inclination for a random distribution of rotation axes orientations. One can see that all three dependences are very close to each other. Noticeable deviations in the inflection point position happen only for faculae-dominated stars with inclination value of 45° (red asterisks in the right part of Fig. 2.9).

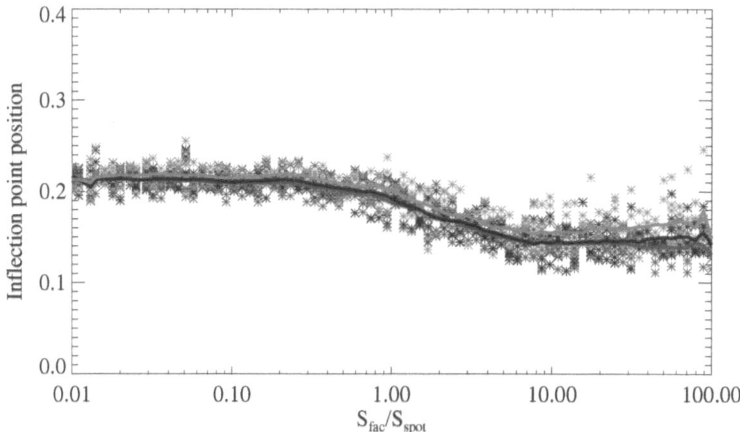

Figure 2.9: Sensitivity of inflection point to inclination of stellar rotation axis. Plotted is the dependence of the inflection point position (given as a fraction of the rotation period, P_{rot} = 30 d) on the ratio between facular and spot disk-area coverage at the time of maximum area, S_{fac}/S_{spot}. Each S_{fac}/S_{spot} value corresponds to five realizations calculated with inclination φ = 90° (equatorial view, blue), φ = 57° (black), and φ = 45° (red). Red, black, and blue lines show positions of the inflection points averaged over the five corresponding realizations.

2.6   Position of the inflection point as a function of stellar magnetic activity

The main goal of this section is to connect the position of the inflection point with proxies of stellar magnetic variability, namely with the S-index and photometric variability. When the facular to spot area ratio *at the time of maximum area*, S_{fac}/S_{spot}, is fixed, the position of the inflection point does not show any dependence on the total coverage of stellar surface by active regions (see Fig. 2.14). At the same time the level of magnetic activity affects the relative areas of facular and spot parts of active regions (Shapiro et al. 2014a) and, consequently, the value of S_{fac}/S_{spot}. This leads to the dependence of the inflection point position on the magnetic activity which we study in this section.

In this context, we have simulated light curves with a different number of active regions emerging on each underlying star over the 1600-day period of simulations. We start with 80 emergence for the "quietest" light curves and end with 81000 emergence for the most "active" light curves. The sizes of spot components of active regions have been randomly chosen according to the log-normal distribution from Baumann and Solanki (2005) (see Sect. 2.4.1). For each of the simulations we have calculated the mean value of the spot disk-area coverage and employed Eq. (1) from Shapiro et al. (2014a) to get the corresponding value of the S-index. Next we employed Eq. (2) from Shapiro et al. (2014a) to obtain the value of the facular disk-area coverage from the S-index. We have corrected this value by subtracting facular coverage corresponding to the absence of spots (0.5% according to Eqs. (1–2) from Shapiro et al. 2014a). Then we have calculated S_{fac}/S_{spot} value, i.e. the ratio at the peak area of the active region, which would result in such an instantaneous facular disk-area coverage. We have considered the $T_{fac}/T_{spot} = 3$ case and set the decay rate of spots to 10 MSH/day (see Sect. 2.5.1).

Resulting dependences of the inflection point position and S_{fac}/S_{spot} value on the S-index are given in the upper panel of Fig. 2.10. One can see that the S_{fac}/S_{spot} value decreases with the S-index. This is because Eqs. (1–2) in Shapiro et al. (2014a) are based on the extrapolation from the solar case, where spot disk-area coverage depends on the S-index quadratically, while the dependence of facular disk-area coverage is linear. A decrease of the S_{fac}/S_{spot} value with the S-index causes a rather weak shift of the inflection point to higher frequencies. For example, one can see that the position of the inflection point slightly shifts from solar minimum to solar maximum. At the same time the shift is smaller than the fluctuations of the inflection point caused by the statistical noise so that it does not contradict the results of Amazo-Gómez et al. (2020b) (see Sect. 2.5.2). Interestingly, the position of the inflection point remains similar to that of the Sun even for significantly more active stars.

For each of the simulated light curves we calculate variability following the definition of variability range by Basri et al. (2011). Namely, we split the light curves into 30-day segments. We sorted the segments by brightness and calculated the range between the 5th and 95th percentile of the brightness. Then we calculated the mean range among all 30-day segments. The resulting variability values are plotted in the middle panel of Fig. 2.10 as a function of the S-index. One can see that although the spot disk-area coverage increases quadratically with the S-index, the increase of the photometric variability is almost linear. This is because the variability range depends not on the absolute value of stellar disk-area coverage by active regions but rather on its fluctuations with time. The rise in the amount of active regions leads to a more uniform surface distribution which, in turn, decreases the variability range.

Middle panel of Fig. 2.10 shows that solar variability range changes from almost zero during the solar minimum to roughly 1.5 ppt (parts per thousand). This agrees with a more accurate calculation in Shapiro et al. (2016) (see their Fig. 10a.). In the lower panel of Fig. 2.10 we plot the dependence of the inflection point position on the variability range. In most of the cases the position of the inflection point remains in between roughly 14% and 21% of the rotation period, even for stars significantly more variable than the Sun.

Lower and upper panels of Fig. 2.10 hint at a seemingly simple way of eliminating the uncertainty in calibration between the stellar rotation period and inflection point position brought about by the unknown facular contribution to stellar variability (see Sect. 2.5.1). One can either estimate the calibration factor from the value of the S-index (if known) or from the amplitude of the photometric variability. However, all the dependences plotted in Fig. 2.10 are produced for a fixed values of the spot decay rate and ratio between facular and spot lifetimes (see above). Both these values are rather uncertain even for the solar case. To take it into account we recalculated all the dependences for a broad range of spot decay rates and ratios between facular and spot lifetimes and plotted them in Fig. 2.11. One can see that the resulting dependences are noisier than those plotted in Fig. 2.10. This is because a) the spot decay rate affects the connection between the number of emergence and *instantaneous* spot disk-area coverage (which defines the value of the S-index) and

2 Inflection point in the power spectrum of stellar brightness variations: I. The model

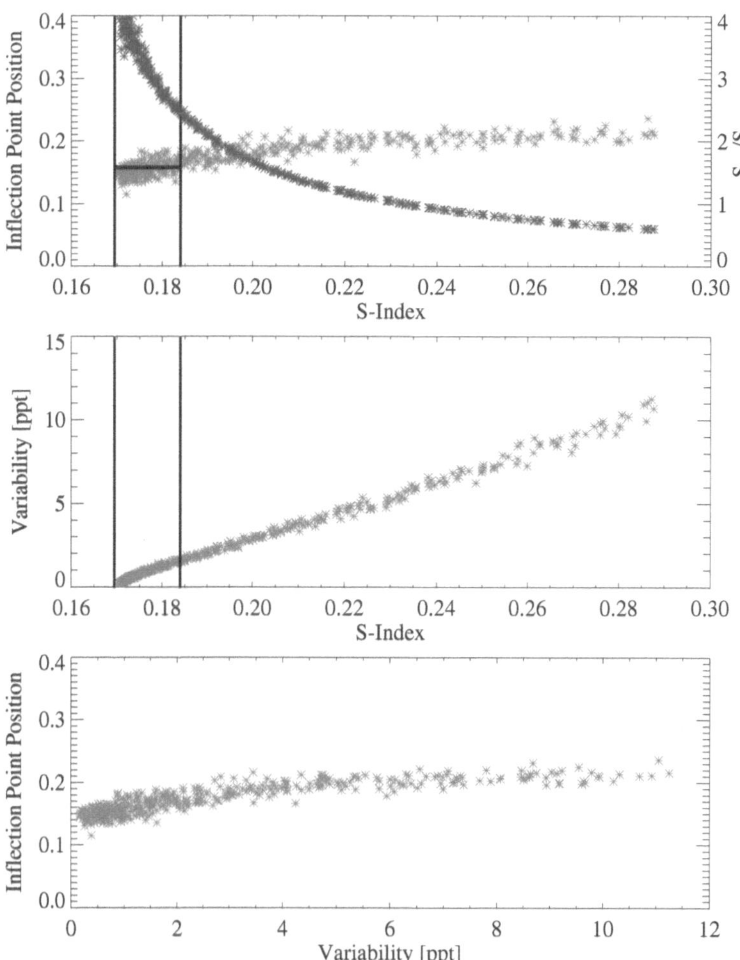

Figure 2.10: Dependence of the inflection point position (given as a fraction of the rotation period, $P_{\rm rot} = 30$ d) on the S-index and stellar photometric variability (shown in red in top and bottom panels, respectively) as well as the dependence of the photometric variability on the S-index (middle panel). Blue asterisks in upper panel indicate the dependence of the facular to spot ratio at the time of maximum area, $S_{\rm fac}/S_{\rm spot}$, on the S-index. Black vertical lines in upper and middle panels point to the range of solar S-index values, while the horizontal black line in the top panel corresponds to the position of solar inflection point from Amazo-Gómez et al. (2020b).

b) the ratio between facular and spot lifetimes is in charge of the connection between *instantaneous* disk area coverage and those *at the time of maximum area.*

All in all, Fig. 2.11 shows that, despite a significant level of noise, most of the inflection points for stars with variability ranges below 3 ppt are located in between 13% (for faculae-dominated stars) and 21% (for spot-dominated stars) of the rotation period. In this respect, we suggest that the best algorithm for determining rotation periods of stars similar to the Sun would be to take a solar value of about 16% (solar value, see Sect. 2.5.2), keeping in mind that the intrinsic uncertainty of our method is about 25%.

We must, however, give some words of caution. We assumed that brightness variations of stars with near solar values of effective temperatures can be calculated by a simple extrapolation of the solar model. In other words, we disregarded the potential presence of active longitudes in the emergence of active regions (we note, however, that the existence of active longitudes have been also proposed for the Sun, see e.g. Berdyugina and Usoskin 2003), and we assumed a solar distribution of sizes of active regions, solar spot decay rates, as well as solar ratios between facular, spot umbra, and spot penumbra areas.

The presence of active longitudes might significantly amplify the amplitude of brightness variations and simultaneously make the rotation peak in the power spectra more pronounced. Along the same line, while we do not expect that the size distribution of active regions has a direct impact on the position of the inflection point, it can influence the photometric variability and hence affect the dependence plotted in the lower panel of Fig. 2.11. Finally, there is the critical assumption that the dependence of facular and spot disk-coverage on stellar activity (expressed via the S-index) follow the solar relationships. Any deviations from the assumed relationships might affect both the position of the inflection point and the amplitude of the photometric variability. We note, however, that solar relationships proved to be very successful for modeling stellar brightness variations on timescales of magnetic activity cycle (Shapiro et al. 2014a).

2.7   Conclusions

We have developed a physics-based model for calculating stellar brightness variations. The model is loosely based on the highly successful SATIRE approach for modeling solar brightness variations.

We have utilized our model to show that the rotation signal in the photometric records of stars with near solar fundamental parameters and rotation periods is significantly weakened by a) short lifetimes of spots; b) partial compensation of spot and facular contributions to the rotation signal. Both these factors can also lead to the appearance of "rogue" global maxima in the power spectra of stellar brightness variations. These maxima are not associated with the rotation period and can mislead the standard methods for rotation period determination. We construe this as the explanation for the low success rates in detecting rotation periods of stars similar to the Sun (van Saders et al. 2019; Reinhold et al. 2019).

We have shown that even in the absence of the rotation peak in the power spectra of stellar brightness variations the information about the rotation period is still contained in the high-frequency tail of the power spectrum. In particular, the rotation period can be determined by applying a pre-calculated calibration factor to the frequency corresponding

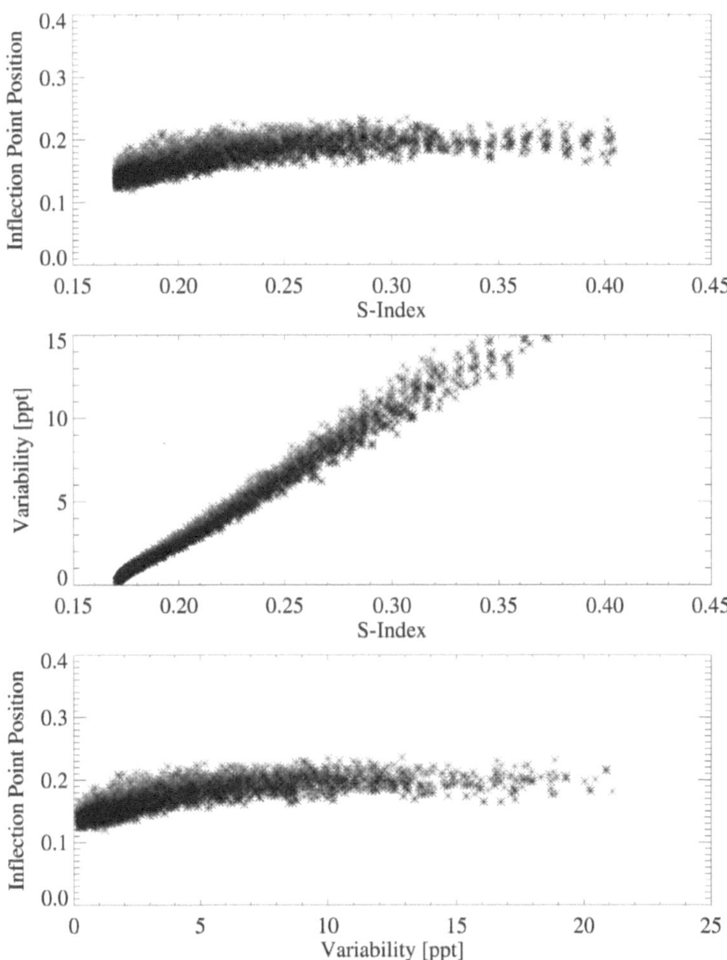

Figure 2.11: The same dependences as in Fig. 2.10 but computed for nine pairs of spot decay rates and ratios between facular and spot lifetimes. For each pair of these parameters, calculations are performed for five realizations of active regions emergence. Shown are calculations with $T_{\text{fac}}/T_{\text{spot}} = 2$ (black), $T_{\text{fac}}/T_{\text{spot}} = 3$ (blue), and $T_{\text{fac}}/T_{\text{spot}} = 5$ (red). For each $T_{\text{fac}}/T_{\text{spot}}$ ratio we perform calculations with three values of spot decay rates: 41 MSH/day, 25 MSH/day, and 10 MSH/day.

to the inflection point, i.e. the point where the concavity of the power spectrum plotted in the log-log scale changes sign. We have demonstrated that the calibration factor only weakly depends on the parameters describing the evolution of stellar active regions (e.g. their lifetime), stellar disk-area coverage by active regions, and stellar inclination. At the same time the calibration factor depends on the relative areas covered by spots and faculae. On the one hand, this introduces intrinsic uncertainty in the periods determined with our method. On the other hand, the dependence of the calibration factor on the ratio between facular and spot-area coverage allows measuring this ratio in stars with known rotation periods. This might be interesting for constraining the properties of flux emergence in Sun-like stars (see, e.g. Işık et al. 2018).

We have shown that the ratio between the inflection point position and rotation period is about 0.2–0.23 for the purely spot-dominated stars, which are supposedly much more active than the Sun (see, e.g. Shapiro et al. 2014a). The presence of faculae decreases the ratio so that we expect it to be in between 0.13 and 0.21 for stars with near-solar level of photometric variability. Despite a significant uncertainty the main advantage of our method is that it can be used for determining rotational periods of stars with irregular light curves where other available method fails. For such stars we recommend to use solar value of the ratio, i.e. 0.16, which should return rotational period with roughly 25% uncertainty.

We intend to further develop the model presented in this study as well as to apply it to available stellar photometric data. On the theoretical side we plan to a) extend the present study to stars with various fundamental parameters by replacing the Unruh et al. (1999) spectra of the quiet Sun and solar magnetic features with recent simulations of stellar spectra (see, e.g. Beeck et al. 2015; Norris et al. 2017; Witzke et al. 2018); b) utilize recent simulations of magnetic flux emergence and transport by Işık et al. (2018) to better describe the evolution of active regions.

On the observational side we plan to a) test our method for the determination of the rotation period against available solar photometric data (see, Amazo-Gómez et al. 2020b) and stars with known rotation periods; b) apply our method to the sample of *Kepler* (and, in future, TESS) stars with unknown rotation periods.

Acknowledgements chapter 2

We thank Robert Cameron and Yvonne Unruh for useful discussions. The research leading to this paper has received funding from the European Research Council under the European Union's Horizon 2020 research and innovation program (grant agreement No. 715947). EMAG acknowledges support by the International Max-Planck Research School (IMPRS) for Solar System Science at the University of Göttingen. It also got a financial support from the BK21 plus program through the National Research Foundation (NRF) funded by the Ministry of Education of Korea. We would like to thank the International Space Science Institute, Bern, for their support of science team 446 and the resulting helpful discussions.

2.8   Appendix chapter 2

2.8.1   Additional figures

The section includes Figs. 2.12–2.16.

2.8.2   Solar values of facular to spot area ratio at the time of maximum area

As discussed in Sect. 2.4.1 the directly observed *instantaneous* ratio between solar disk coverage by faculae and spots is different from the ratio between facular and spot areas in individual magnetic regions *at the time of their maximum area*, $S_{\mathrm{fac}}/S_{\mathrm{spot}}$. At the same time in Sect. 2.5 we have shown that the position of the inflection point is defined by the ratio *at the time of maximum area*, $S_{\mathrm{fac}}/S_{\mathrm{spot}}$ and thus we need to know solar value of $S_{\mathrm{fac}}/S_{\mathrm{spot}}$ to check whether position of the inflection point in the power spectrum of solar brightness variations is consistent with the model presented here. In this section we show how to determine solar value of $S_{\mathrm{fac}}/S_{\mathrm{spot}}$ from the observed *instantaneous* records of solar disk coverage by spots and faculae (see Sect. 2.5.2).

First we employ the model setup described in Sect. 2.4.1 to calculate the power spectra of modeled facular and spot disk area coverage as they would be seen along the stellar rotation axis and from the stellar equatorial plane. We consider $P_{\mathrm{rot}} = 30$ d, $S_{\mathrm{fac}}/S_{\mathrm{spot}} = 3.4$, $T_{\mathrm{fac}}/T_{\mathrm{spot}} = 3$ case, adopt log-normal distribution of spot sizes from Baumann and Solanki (2005) and 25 MSH/day for the sunspot decay rate.

Left panels of Fig. 2.17 show the global wavelet (Morlet, 6th order) power spectra (top) of *instantaneous* disk coverage by spots and faculae observed along rotation axis of a modeled star (so that the rotational modulation does not affect the power spectra) as well as their ratio (bottom). One can see that the ratio is roughly constant and is equal to $S_{\mathrm{fac}}^2/S_{\mathrm{spot}}^2$ up to the period of about 90 days (i.e. $3P_{\mathrm{rot}}$). This is not surprising since the decay of magnetic features only affects the power spectrum at timescales larger than the decay time.

More strictly, when observing along the rotation axis, the disk area coverage by the individual magnetic feature is proportional to the product of the unit step function (i.e. function which returns 0 before the emergence of the magnetic feature and 1 after the emergence) and a function describing linear decay. The power spectral density of such a right-triangle function (hereafter function $\mathcal{F}_1(t)$) is proportional to:

$$\mathcal{D}(\nu) \sim \frac{S_{\mathrm{feature}}^2}{x^2} + \frac{S_{\mathrm{feature}}^2}{x^4}\left[\sin(x/2)^2 + \sin((x-\pi/2)/2)^2\right], \tag{2.5}$$

where S_{feature} is the disk area coverage of the feature *at the time of maximum area*, $x \equiv 2\pi T_{\mathrm{dec}}\nu$, and T_{dec} is a decay time of magnetic feature (see Eq. 2.2).

At $x \gg 1$ (which corresponds to $P \ll 2\pi T_{\mathrm{dec}}$, where $P \equiv 1/\nu$) the second term on the right-hand side of Eq. 2.5 becomes negligibly small in comparison to the first term. Consequently, the corresponding power spectral density $\mathcal{D}(\nu)$ drops with frequency as $1/\nu^2$ independently of the decay time of magnetic feature. Hence, the ratio between power spectra of facular and spot disk area coverage of an active region (consisting of facular and spot features, see Sect. 2.4) is equal to $S_{\mathrm{fac}}^2/S_{\mathrm{spot}}^2$. We note that the power

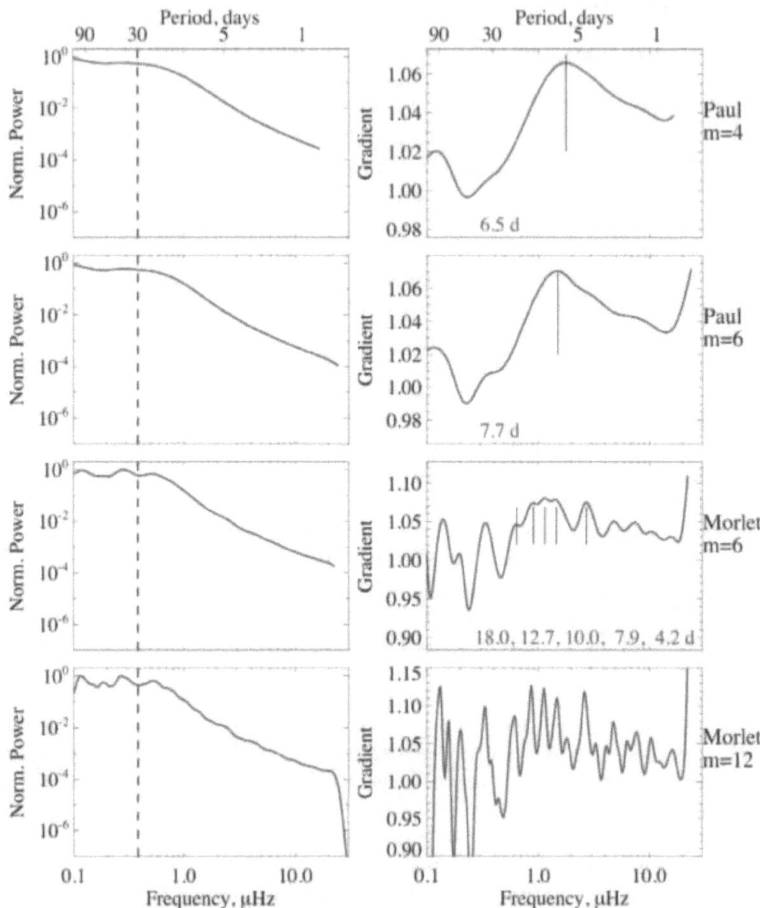

Figure 2.12: Global wavelet power spectra (left panels) of the light curve from Fig. 2.1 and corresponding gradients of the power spectra (right panels) calculated with different wavelets. The frequency localization of the utilized wavelet is increasing from the top to bottom panels. As in Fig. 2.1 vertical dashed lines in the left panels indicate the rotation period of the modelled star. Vertical solid lines and numbers in the right panels (not shown in the bottom panel) indicate positions of the inflection points.

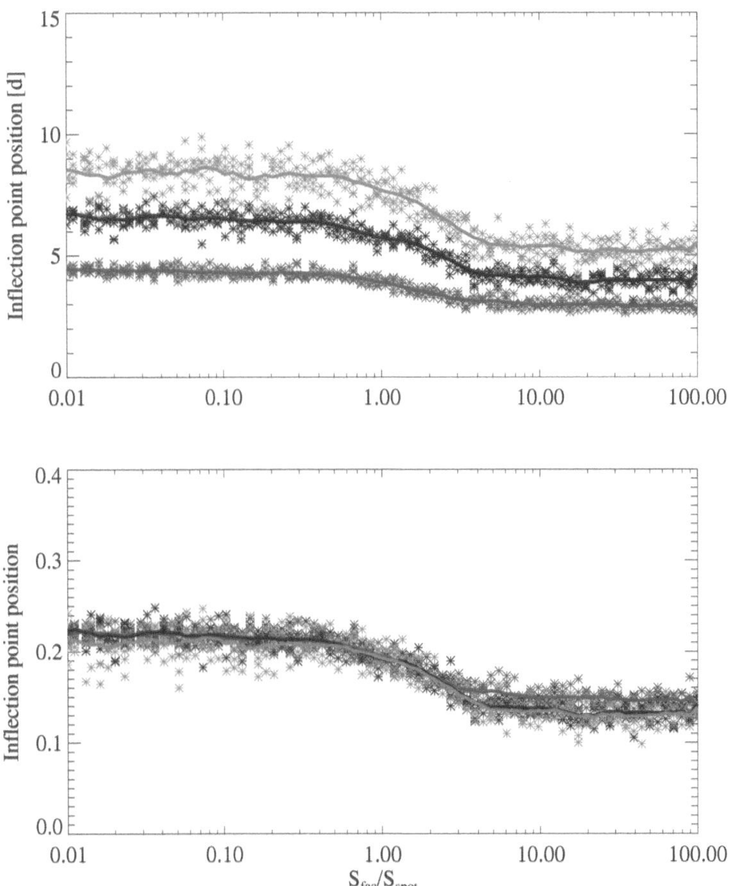

Figure 2.13: Dependence of the inflection point position on the ratio between facular and spot disk-area coverage at the time of maximum area, S_{fac}/S_{spot}, plotted for three values of the rotation period: 20 d (blue), 30 d (black), and 40 days (red). Shown are positions in days (upper panel) and ratios with respect to the rotation period (lower panel). Calculations are performed for a spot decay rate of 25 MSH/day, $T_{fac}/T_{spot} = 3$, and mean fractional disk-area spot coverage of 0.3%. As in Figs. 2.8–2.9 for each pair of S_{fac}/S_{spot} and rotation periods values, five realization of active regions emergence are shown. Red, black, and blue lines indicate positions of the inflection points averaged over corresponding five realizations. A small deviation of 20 d curve from 30 d and 40 d curves in the lower panel at high S_{fac}/S_{spot} values can be explained by the insufficient cadence of light curves (4 points per day) for 20 d rotation period and aliasing effect.

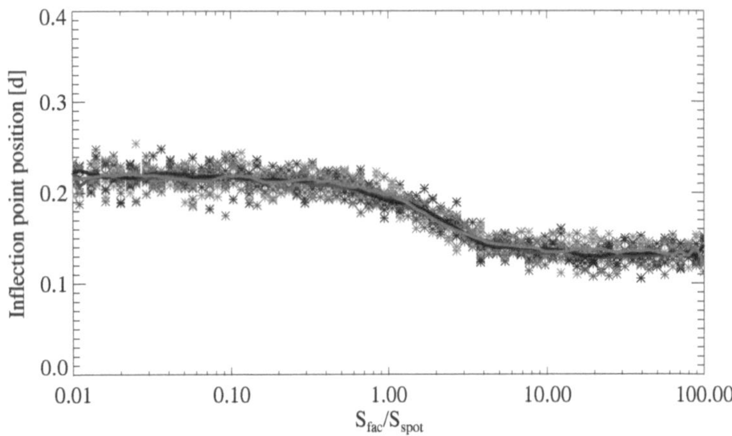

Figure 2.14: Same as Fig. 8, but for three values of mean fractional disk-area spot coverage: 0.075% (blue), 0.3% (black), and 0.75% (red). Calculations are performed for spot decay rate of 25 MSH/day, $T_{\mathrm{fac}}/T_{\mathrm{spot}} = 3$ and rotation period of 30 d.

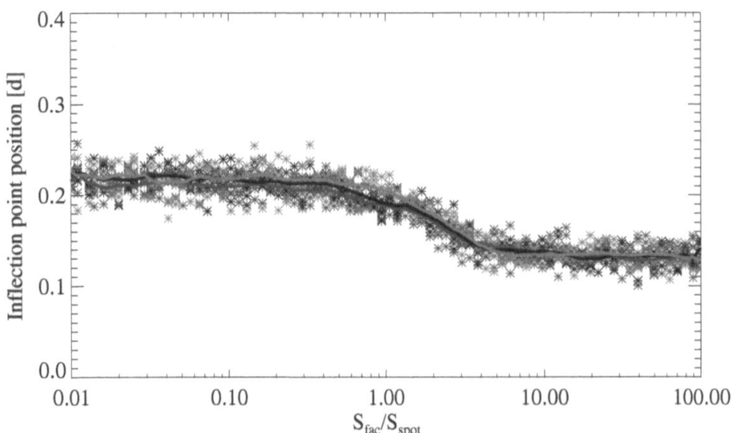

Figure 2.15: Same as Fig. 8, but for three values of spot decay rate: 10 MSH/day (blue), 25 MSH/day (black), 41 MSH/day (red). Calculations are performed for mean fractional disk-area spot coverage of 0.3%, $T_{\mathrm{fac}}/T_{\mathrm{spot}} = 3$ and rotation period of 30 d.

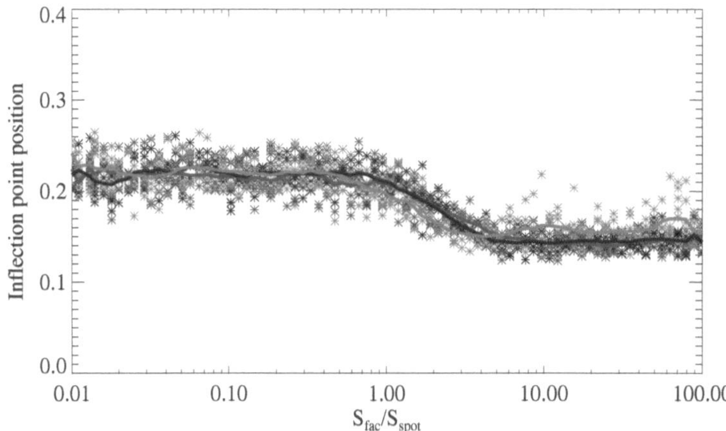

Figure 2.16: Same as Fig. 8, but calculated assuming that all emergence of active regions happen at the equator (blue), latitudes $\pm 30°$ (black), and latitudes $\pm 60°$ (red). Calculations are performed for mean fractional disk-area spot coverage of 0.3%, $T_{\mathrm{fac}}/T_{\mathrm{spot}} = 3$ and rotation period of 30 d.

spectra of the observed *instantaneous* facular and spot disk coverage are brought about by the superposition of the contributions from many incoherently emerging active regions. Therefore the ratio of high-frequency parts of the power spectra represents the mean $S_{\mathrm{fac}}/S_{\mathrm{spot}}$ value over all active regions. Since the contribution of magnetic features to the disk coverage is proportional to their size this mean value is weighted towards larger active regions.

Let us now consider the case of the observations from the stellar equatorial plane (i.e. the solar case since the $\approx 7.25°$ angle between the solar equator and ecliptic can be neglected in our analysis). Middle panels of Fig. 2.17 show the global wavelet power spectra of facular and spot disk coverage resulting from the same realization of active regions emergence as shown in the left panels, but now the active regions are observed from the equatorial plane of the modeled star. One can see that the ratio between facular and spot power spectra is strongly affected by rotation at periods below 45 d (i.e. $3/2\,P_{\mathrm{rot}}$) but is basically not affected by the rotation at periods between 45 d and 90 d. Below we give an explanation of such a behavior.

The disk area coverage by a single transiting magnetic feature for a star observed from the equatorial plane is proportional to the product of three functions: a. the right-triangle function $\mathcal{F}_1(t)$ (with power spectral density given by Eq. 2.5); b. a function which returns zero during half of the period when the feature is at the far-side of the star and 1 during another half of the period when the feature is on the near-side of the star (i.e. shifted by 0.5 square wave function, hereafter function $\mathcal{F}_2(t)$); c. function describing foreshortening effect (hereafter function $\mathcal{F}_3(t)$).

The Fourier transform of the square wave function contains only odd-integer harmonics of the form $\pm(2k-1)v_{rot}$ where $v_{rot} = 1/P_{rot}$. The shift by 0.5 brings about an additional zero frequency component so that the Fourier transform of function $\mathcal{F}_2(t)$ contains zero frequency and odd-integer harmonics. The foreshortening function $\mathcal{F}_3(t)$ is proportional to the cosine of the angle between the direction from the centre of the star to the observer and to the magnetic feature and contains only $\pm v_{rot}$ components. Multiplication in the time domain corresponds to the convolution in the frequency domain. Consequently, the Fourier transform of the product of $\mathcal{F}_2(t)$ and $\mathcal{F}_3(t)$ functions contains all harmonics of the rotational period $\pm k v_{rot}$ and zero frequency component.

All in all, the disk area coverage observed from the stellar equatorial plane can be obtained by multiplying disk area coverage observed along the stellar rotation axis (which are proportional to $\mathcal{F}_1(t)$) with a function containing zero frequency component and harmonics of the rotational frequency. As discussed above, the power spectra of facular and spot disk coverage are proportional to each other with the exception of the $[0, v_{rot}/3]$ interval. After the convolution the signal in this interval will be propagated to $[| \pm k\,v_{rot} |, | v_{rot}/3 \pm v_{rot} |]$ (i.e. $[0, v_{rot}/3]$, $[v_{rot}, 4/3\,v_{rot}]$, $[2\,v_{rot}, 7/3\,v_{rot}]$..., and $[2/3\,v_{rot}, v_{rot}]$, $[5/3\,v_{rot}, 2\,v_{rot}]$, ...) intervals. Interestingly, the signal does not propagate to the $[1/3\,v_{rot}, 2/3\,v_{rot}]$ (or $[3/2\,P_{rot}, 3\,P_{rot}]$) interval. This explains the curious behavior of the ratio between power spectral density of facular and spot disk coverage in this interval shown in the middle lower panel of Fig. 2.17: neither the decay of magnetic features nor the stellar rotation affects it.

The power spectra and their ratio shown in right panels of Fig. 2.17 are calculated using solar disk area coverage obtained by Yeo et al. (2014) using solar magnetograms and continuum images recorded by the Helioseismic and Magnetic Imager onboard the Solar Dynamics Observatory (SDO/HMI) for the period from May 2010 till August 2014. In agreement with the previous discussion the ratio between the two power spectra is roughly constant in the time interval between 45 d and 90 d and corresponds to S_{fac}/S_{spot} value of 3.

2.8.3   Examples of testing GPS method on *Kepler* stars

An extensive test of our method for determining stellar rotation periods will be in focus of the forthcoming publications. In particular, we will analyze dependence of the inflection point position on the photometric variability and test the dependence established in Sect. 5.2. Here we, as an example, apply our method to stars significantly more variable than the Sun with presumably spot-dominated variability.

In the upper panels of Figs. 2.18–2.20 we show the *Kepler* light curves of KIC2141852 (Fig. 2.18), KIC2553816 (Fig. 2.19), and KIC2992964 (Fig. 2.20). The data for 15 *Kepler* quarters (quarter 2 – quarter 16) have been downloaded from the MAST portal[1] and reduced with the PDC-MAP pipeline (Smith et al. 2012; Stumpe et al. 2014). The lower panels of Figs. 2.18–2.20 show the position of the inflection points for each of the *Kepler* quarters. The behavior of the inflection points is very similar to that shown for the synthesized light curves in Fig. 2.7. Namely, inflection points fluctuate around the mean position, and from time to time "rogue" inflection points, mainly corresponding to the high-period branch, appear. We note that a large number of high-period inflection points imply that the lifetime of magnetic features on considered stars is comparable or larger than their rotational periods (see Sect. 2.3.2). This is consistent with highly regular light curves of

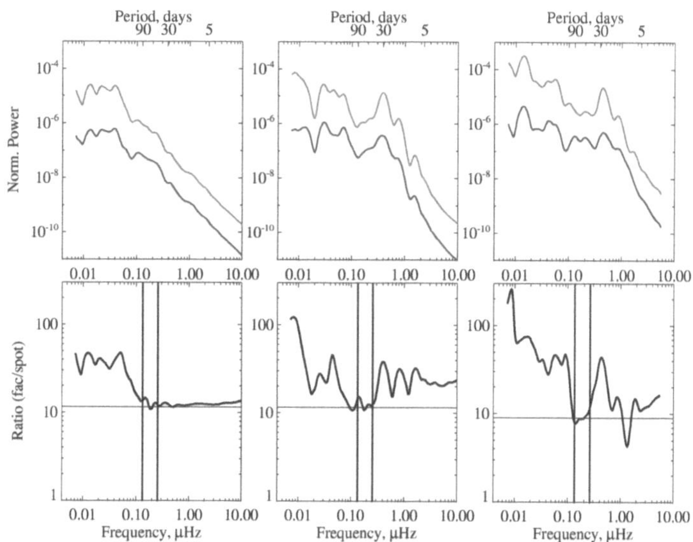

Figure 2.17: Power spectra of facular (red) and spot (blue) disk area coverage for a modeled star observed along its rotational axis (left upper panel), a modeled star observed from its equatorial plane (middle upper panel), and for the Sun deduced from the SDO/HMI observations (right upper panel). The corresponding ratios between the power spectra of facular and spot disk area coverage are plotted in the lower panels. The vertical black lines in the lower panel denote the interval between 45 and 90 days. The horizontal black line in the left and middle lower panel denotes the $S^2_{\mathrm{fac}}/S^2_{\mathrm{spot}}$ value used in the simulations (see text for more details). The horizontal black line in the right lower panels denotes ratio value of 9.

considered stars.

We have calculated the outlier-resistant mean of the low-period inflection point positions as well as the standard error of the mean. Since we expect that the variability of our exemplary stars is spot-dominated, the ratio between the inflection point position and the rotational period should lie between 0.2 and 0.23 (see Fig. 2.8). We applied these factors to the mean position of the inflection point taking the standard error of the mean into account. The resulting ranges for the rotation periods (P_{GPS}) are indicated in Figs. 2.18–2.20 and compared to the periods (P_{R2013}) reported by Reinhold et al. (2013). One can see that periods from Reinhold et al. (2013) are within the range given by our method for all three considered stars.

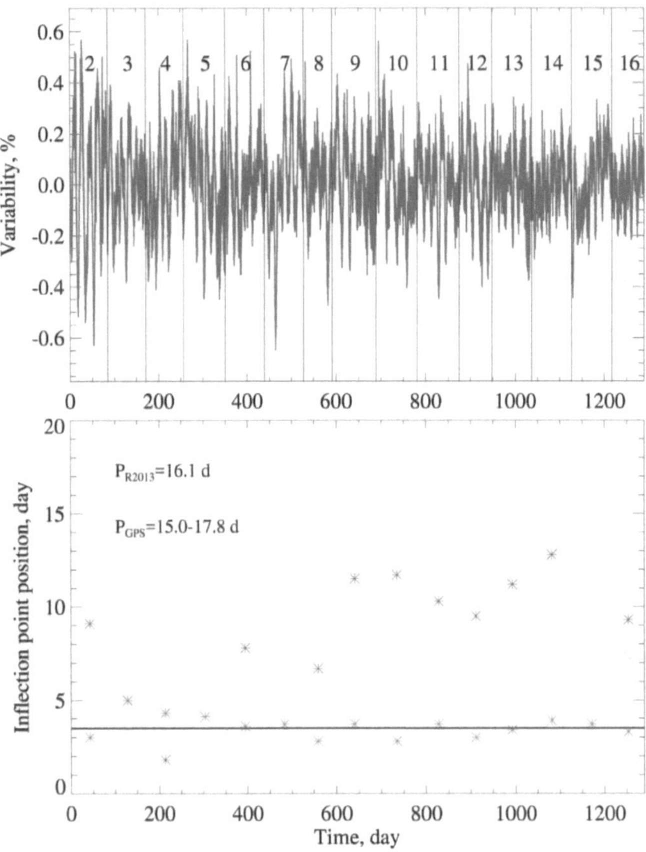

Figure 2.18: The light curve of KIC2141852 (upper panel) and positions of the inflection points for each of the *Kepler* quarters (lower panel). The *Kepler* quarters 2–16 are numbered in the upper panel and separated by the vertical black lines. The asterisks in the lower panel correspond to the positions of inflection points. The value of the outlier-resistant mean of the inflection point positions in the low-period branch (see text for more details) is indicated with the horizontal black line. Red asterisks correspond to the inflection points utilized for calculating the outlier-resistant mean value, blue asterisks are trimmed as outliers. The rotation period range returned by our method (P_{GPS}) and period from Reinhold et al. (2013) (P_{R2013}) are listed in the lower panel.

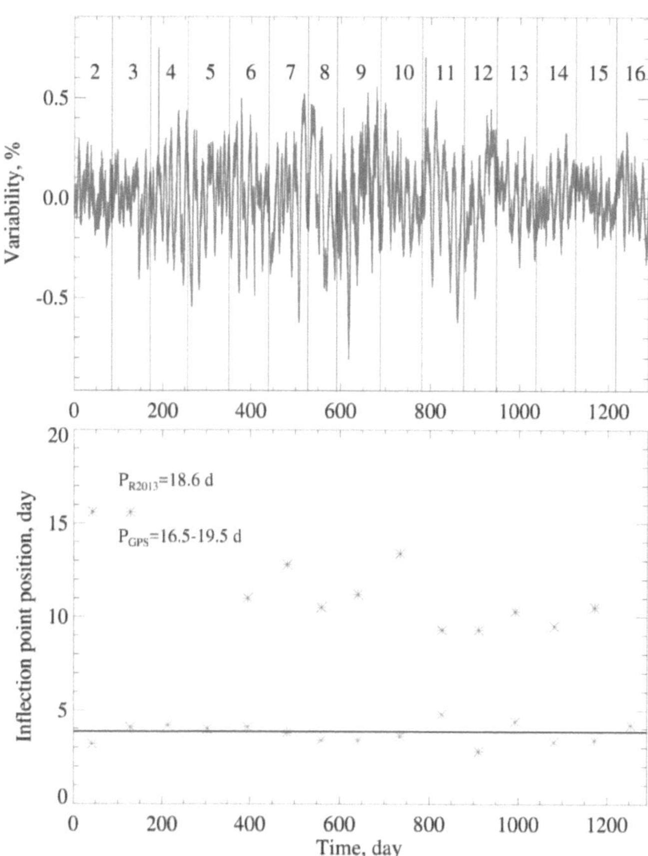

Figure 2.19: The same as 2.18 but for KIC2553816.

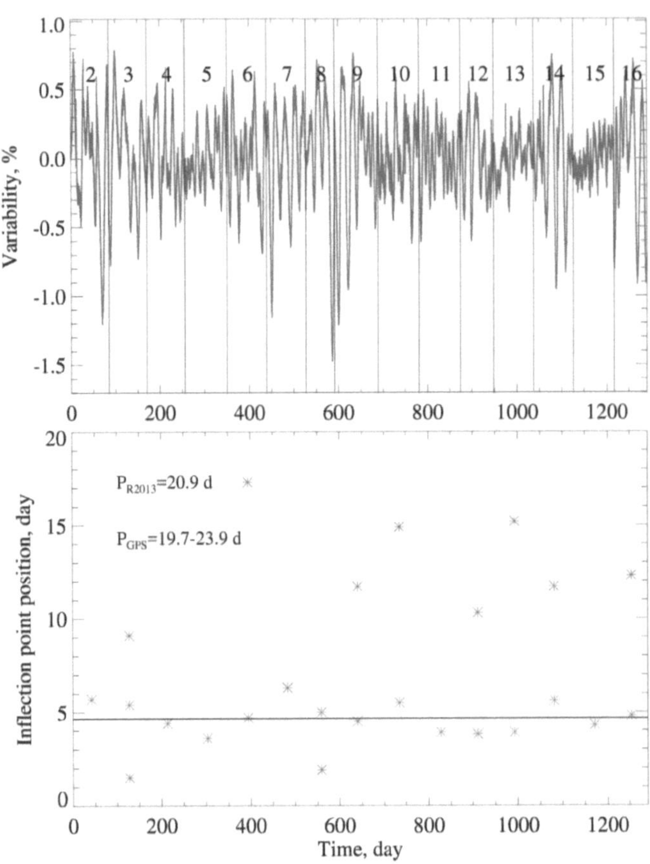

Figure 2.20: The same as 2.18 but for KIC2992964.

3 Inflection point in the power spectrum of stellar brightness variations. II. The Sun

3.1   Introduction of chapter 3

The star's rotation period defines the action of the stellar dynamo, transport of magnetic flux through the convective zone, and its emergence on the stellar surface. (See Charbonneau 2010 for a general review of dynamo theory, and Reiners et al. 2014, Işık et al. 2018 for the relation between rotation period and activity.) Furthermore, for the unsaturated regime (i.e., when Rossby number is larger than 0.13, equivalent to rotation periods longer than 1-10 days for solar-type stars) the magnetic activity of stellar chromospheres (as indicated by the Ca II H & K emission core lines) and coronae (as indicated by the X-ray emission) monotonically increases with the decrease of the rotation period (Pizzolato et al. 2003; Wright et al. 2011; Reiners 2012; Reiners et al. 2014).

Recent studies indicate that the relationships between rotation period, coronal and chromospheric activity work not only for stars with a tachocline (the transition region between the radiative core and convective envelope, see Spiegel and Zahn 1992), but also for slowly-rotating fully convective stars (see Reiners et al. 2012; Wright and Drake 2016; Newton et al. 2017; Wright et al. 2018). The rotation period thus appears to be a good proxy of the overall magnetic activity of a star.

Accurate measurements of rotation periods are important for understanding stellar evolution and for better calibration of the gyrochronology relationship (Ulrich 1986; Barnes 2003). The knowledge of the stellar rotation period helps distinguish the radial velocity jitter of a star from the planetary signal (see, e.g. Dumusque et al. 2011; Oshagh 2018). This is crucial for detection, as well for confirming Earth-size planets in ongoing and upcoming surveys, such as the **ESPRESSO** (see Pepe et al. 2010) and **PLATO** missions (see Roxburgh et al. 2007; Rauer et al. 2016, 2014).

Rotation periods can be determined from photometric observations thanks to the presence of magnetic features on stellar surfaces. Concentrations of strong localised magnetic fields emerge on the stellar surface and lead to the formation of photospheric magnetic features, such as bright faculae and dark spots (Solanki et al. 2006). The transits of these magnetic features over the visible disk as the star rotates imprints particular patterns onto the observed light-curve. These patterns provide a means of tracing stellar rotation periods. We note that in the case of the Sun, such brightness variations are well understood (Solanki et al. 2013; Ermolli et al. 2013) and modern models can explain more than 96% of the variability of total solar irradiance (TSI, which is the spectrally-integrated solar radiative flux at one Astronomical Unit from the Sun) (see Yeo et al. 2014).

Available methods for retrieving rotation periods from photometric time series, e.g., autocorrelation analysis or Lomb-Scargle periodograms, appeared to be very successful for determining periods of active stars with periodic patterns of variability. The transiting planet-hunting missions such as COROT, Kepler, and TESS, (Bordé et al. 2003a; Borucki et al. 2010; Ricker et al. 2015) opened unprecedented possibilities for acquiring accurate high-precision photometric time-series of stars different from the Sun. The new data from these missions enabled studies of stellar magnetic activity on a completely new level. In particular, it has become possible to measure rotation periods for tens of thousands of stars (see, Walkowicz and Basri 2013; Reinhold et al. 2013; Nielsen et al. 2013; García et al. 2014; McQuillan et al. 2014; Buzasi et al. 2016a; Angus et al. 2018). At the same time, the pattern of brightness variations in slow rotators such as the Sun is often quasi-periodic and even irregular. The irregularities are mainly caused by the short (in comparison to stellar-rotation period) lifetimes of magnetic features, such as starspots/sunspots, and a large degree of randomness in the time and position of their emergence on the stellar surface. This renders the determination of rotation periods for low activity stars very difficult. For example, van Saders et al. (2019) showed that rotation periods of about 80% of stars in the *Kepler* field of view with near-solar effective temperature remain undetected. Consequently, the stars with known rotation periods represent only the tip of the iceberg of Sun-like stars. This can lead to biases in conclusions drawn based on the available surveys of stellar rotation periods. The relatively low efficiency of standard methods in detecting periods of stars with non-regular light curves might affect studies aimed at comparisons of solar and stellar variability. For example, solar variability appears to be normal when compared to main-sequence *Kepler* stars with near-solar effective temperatures (Basri et al. 2013; Reinhold et al. 2020a). At the same time, when comparisons are limited to main-sequence stars with near-solar effective temperature and with *known* near-solar rotation periods, the solar variability appears to be anomalously low (Reinhold et al. 2020a). One possible explanation of such a paradox is the inability of standard methods to reliably detect rotation periods of stars with light curves similar to that of the Sun (see also discussion in Witzke et al. 2020). Along the same line, Reinhold et al. (2019) proposed that biases in determining rotation periods might contribute to the explanation of a dearth of intermediate rotation periods observed in *Kepler* stars (McQuillan et al. 2013; Reinhold et al. 2013; McQuillan et al. 2014; Davenport 2017).

In Shapiro et al. (2020) (hereafter, Paper I), we proposed a new method for determining stellar rotation periods from the records of their photometric variability. The method is applicable to late-type stars but particularly beneficial for and aimed at stars with low activity and quasi-periodic irregularities in their light-curves. The rotation period is determined from the profile of the high-frequency tail (i.e., its part in between about a day and 5–10 days) of the smoothed wavelet power spectrum. For this work we used Paul wavelet of order six (see Torrence and Compo 1998). We identified the point where the gradient of the power spectrum (GPS) plotted on a log-log scale (in other words, $d\,(\ln P(\nu))/d(\ln \nu)$, where P is power spectral density and ν is frequency) reaches its maximum value. This point corresponds to the inflection point, i.e., the concavity of the power spectrum plotted in the log-log scale changes sign there. In Paper I we have shown that the period corresponding to the inflection point is connected to the rotation period of a star by a calibration factor which is a function of stellar effective temperature, metallicity, and activity.

The main goal of the present study is to test and validate the method proposed in Paper I against the Sun, which presents a perfect example of a star with an irregular pattern of variability but known rotation period. In particular, by validating the GPS method against the Sun, we show that it has the potential to reduce biases caused by the relative inefficiency of standard methods to determine rotation periods of low-activity stars. More specifically, we utilise the calibration factor between the position of the inflection point and rotation period corresponding to the solar case and apply the GPS method to the available photometric records of the Sun. Furthermore, we test the performance of the GPS method at various levels of solar activity.

In Sect. 3.2, we give a short overview of the available methods for determining rotation periods as well briefly describe the GPS method. (A more detailed discussion of this method is given in Paper I.) In Sect. 3.3, we compare the performance of our method with that of other available methods in the exemplary case of the Sun. In Sect. 3.4, we present the relationship between solar activity and skewness of its photometric light-curve. Our main results are summarised in Sect. 3.5.

3.2   Methods for determining stellar-rotation periods

High-precision, high-cadence photometric time-series allow the determination of accurate stellar-rotation periods, and the scientific community has developed various methods for retrieving periodicities embedded in these data. Current methods include autocorrelation functions, Lomb-Scargle periodograms, wavelet power spectra, and very recent techniques based on the Gaussian processes.

The autocorrelation function (ACF) method is based on the estimation of a degree of self-similarity in the light-curve over time. The time lags at which the degree of self-similarity peaks are assumed to correspond to the stellar-rotation period and its integer multiplets. This assumption is valid if magnetic features which cause photometric variability are stable over the stellar-rotation period.

The ACF method has been used by McQuillan et al. (2014) to create the largest available catalog of rotation periods until now: the rotation periods were found for 34,030 (25.6%) of the 133,030 main-sequence *Kepler* target stars (observed in the *Kepler* star field at the time of the publication). The ACF calculations presented in our study have been performed with the A_CORRELATE IDL function.

The Lomb-Scargle periodogram gives the the power of a signal a certain frequency (see Lomb 1976; Scargle 1982). Here we use the generalised Lomb-Scargle (GLS) version v1.03, applying the formalism given by Zechmeister and Kürster (2009). The GLS method is widely used for retrieving periodicities from time-series, and is applicable to stellar light-curves with non-regular sampling. In studies aimed at the determination of the rotation period, the highest peak in the GLS is assumed to correspond to the rotation period (see Reinhold et al. 2013).

Wavelet power spectra (hereafter, PS) transform analysis is beneficial for time-series that have a non-stationary signal at many different frequencies. It has been employed for determining stellar-rotation periods by García et al. (2009). To perform PS, we use the WV_CWT IDL function based on the 6th order Paul wavelet, (see Farge 1992; Torrence and Compo 1998).

Another technique, which is currently being actively developed, is inferring rotation periods with the help of Gaussian processes (GP, Roberts et al. 2012). GP can be applied to detect a non-sinusoidal and quasi-periodic behaviour of the signal in light-curves. The GP regression has been extensively tested for analysing various time-series, and, in particular, for retrieving the periodic modulation from stellar radial-velocity (Rajpaul et al. 2015) and photometric (Angus et al. 2018) signals. GP performance can be readily compared with other approaches for determining rotation periods in small stellar samples (see, e.g., Faria et al. 2020, who applied various techniques to HD 41284). GP calculations, however, demand significant computational resources, e.g., analysis of a typical *Kepler* light curve takes from several hours to 12 hours and longer (Angus et al. 2018). The time efficiency of the GP algorithms is expected to significantly improve with the development of new methods (Foreman-Mackey et al. 2017a,b). Thus, it might be interesting in future studies to compare rotation periods determined by the GP and GPS algorithms. Such a comparison is, however, out of the scope of the present study.

The methods described above have been very successful for determining rotation periods of active stars with periodic patterns of photometric variability. However, their performance deteriorates for slower-rotating and consequently less-active stars, and, in particular, for stars with a near-solar level of magnetic activity. In such stars, the pattern of variability is complex and non-regular due to the irregular emergence of magnetic features that manifest as spots and faculae. In particular, the dark spots can lead to prominent dips in brightness, and we expect that for stars with solar-activity levels, their decay time is comparable to or even less than the stellar-rotation period, as it is for the Sun (see Solanki et al. 2006; Shapiro et al. 2017). Furthermore, Shapiro et al. (2017) showed that in the solar case, the superposition of bright facular and dark spot signatures might lead to a disappearance of the rotation peak from the power spectrum of solar brightness variations. This agrees with many studies (see Aigrain et al. 2015; He et al. 2015, 2018) which have found that the rotation period of the Sun can be reliably determined only during the periods of low solar activity. We note that during these periods, the solar variability is caused by long-lived facular features, and the spot contribution is small.

Recently, He et al. (2015, 2018) analysed solar and stellar brightness using GLS and introduced two indicators: one describing the degree of periodicity in the light-curve, and the other describing the amplitude of the photometric variability. They found that light-curve periodicities of the *Kepler* stars are generally stronger during high-activity times. In contrast, solar light-curves are more periodic during phases of low activity. By applying GLS to the TSI time-series, they could determine the solar-rotation period only during periods of low solar activity.

In Paper I, we employed the SATIRE approach (where SATIRE stands for Spectral and Total Irradiance Reconstruction, see Fligge et al. 2000b; Krivova et al. 2011) to simulate light-curves of stars with solar effective temperature and metallicity for various cases of magnetic-feature emergence and evolution. We found that the profile of the wavelet power spectrum around the rotation period strongly depends on the decay time of magnetic features and the ratio between coverage of the disk by faculae and spots. At the same time, the high-frequency tail of the power spectrum, particularly the portion with periods between about a day and a quarter of the rotation period, remains much more stable, and mainly depends on the rotation period. To quantitatively characterise the high-frequency tail of the power spectrum, we calculated the ratios R_k between the power spectral density, $P(v)$, at two adjacent frequency grid points: $R_k \equiv P(v_{k+1})/P(v_k)$. For a frequency grid equidistant in the logarithmic scale (i.e., with a constant value of $\Delta v/v$), this ratio represents the gradient of the power spectrum plotted on a log-log scale. Consequently, the maxima of the R_k values correspond to the positions of the inflection points, where the concavity of the power spectrum plotted on a log-log scale changes sign. Hereafter, following Paper I, we will refer to the R_k values as the gradient of the power spectrum, GPS.

In Paper I, we found that the position of the high-frequency inflection point (i.e., the inflection point (IP) with highest frequency, see discussion below) is related to the stellar-rotation period by a certain calibration factor defined as: $\alpha = HFIP/P_{\mathrm{rot}}$, where HFIP is a period corresponding to the high-frequency inflection point and P_{rot} is the rotation period. This allowed us to propose a new method for determining the rotation period: calculate the position of the inflection point and scale it with an appropriate value for the factor α to determine the rotation period.

The choice of the scaling factor α is one of the most delicate steps in the GPS method and also one of the main sources of its uncertainty. In Paper I, we found that the value of α is fairly insensitive to the parameters describing the evolution of magnetic features, and, in particular, to the decay time of spots. It shows a stronger dependence on the area ratio between the facular and spot components of active regions at the time of maximum area, S_{fac}/S_{spot}. We have estimated that over the 2010-2014 period, the solar S_{fac}/S_{spot} value was about 3 (see Appendix B of Paper I). For the calculations presented in this paper we adopt $\alpha_{Sun} \pm 2\sigma = 0.158 \pm 0.014$, corresponding to $S_{fac}/S_{spot} = 3$ and the spot decay rate of 25 MSH/day. Such a value of spot decay time is intermediate between the 10 MSH/day estimate by Waldmeier (1955) and 41 MSH/day estimate by Martínez Pillet et al. (1993).

The uncertainty of the α value is computed in Paper I as the standard deviation between positions of inflection points corresponding to different realisations of spot emergence (but the same values of the S_{fac}/S_{spot} ratio and spot decay rates). We stress that when applied to other stars, the uncertainty of our method will be significantly larger since neither S_{fac}/S_{spot} value nor spot decay rates are known a priory. For example, in Paper I we estimated that the internal uncertainty of our method is 25%. Its main advantage, however, is that, in contrast to other methods, it is applicable to inactive stars with irregular patterns of brightness variability.

In summary, we calculate the rotation period and its uncertainty as:

$$P_{\text{rot}} \pm 2\sigma_P = \frac{HFIP}{(\alpha_{Sun} \pm 2\sigma_\alpha)}, \tag{3.1}$$

where HFIP represents the value of the high-frequency inflection point in the power spectrum of brightness variations.

3.3   Validation of GPS method for the solar case

In the present work, we evaluate the GPS method for determining solar rotation period against available records of its brightness variation. We consider the performance of the method at different levels of solar activity using two TSI data sets.

3.3.1   Records of total solar irradiance, TSI

Driven by the interest from the climate community, the solar irradiance has been measured by various space radiometric instruments and reconstructed with many models, see, e.g., reviews by (Ermolli et al. 2013; Solanki et al. 2013).

Among all available solar-irradiance records, the TSI time-series are the most accurate and possess the longest time coverage (see e.g., review by Kopp 2014). In this context, we have opted for testing the performance of the GPS method against the TSI time-series and focused on what are generally considered to be the two most reliable of the available data sets: one obtained by the Variability of solar IRradiance and Gravity Oscillations (VIRGO) experiment on the ESA/NASA SOlar and Heliospheric Observatory (SoHO) Mission (see Fröhlich et al. 1997), and another obtained by the Total Irradiance Monitor (TIM) on the

Solar Radiation and Climate Experiment (SORCE) (see Kopp and Lawrence 2005; Kopp et al. 2005a,b).

VIRGO provides more than 21 years of continuous high-precision, high-stability, and high-accuracy TSI measurements. Our analysis is based on the last update available at the beginning of this work, version 6.4: 6_005_1705, level 2.0 VIRGO/PMO6V observations from January 1996 until June 2017 with a cadence of 1 data point per hour [1]. The data are available via ftp [2].

The TIM data used are version 17, level 3.0, available from February 25, 2003, [1,2]. While TIM data cover a shorter time interval than VIRGO, they have a lower noise level (Kopp 2014), which is particularly important for our analysis of TSI variations during the minimum of solar activity. Here we use an average version of TSI with a cadence of 1 data point per 1.6 hours obtained by averaging over a full orbit of the spacecraft. [3].

3.3.2   Brightness signature of spot transits

In this section we analyse the performance of the GPS method during a 90-day time interval when TSI variability was generated by three consecutive spot transits. Sunspot transits imprint characteristic signals on the TSI, diminishing the observed solar brightness. Transits of well-resolved sunspots have been recorded by TIM/SORCE from December 2006 to February 2007 (see panel (a) of Figure 3.1). One can see a clear signal from spots transiting the solar disk with a recognisable "V-like" shape brought about by the combination of foreshortening effects, the Wilson depression (i.e., spot umbrae are slightly depressed from the photospheric optical-depth unity level due to higher transparency of the spot atmosphere compared to the quiet photosphere (see Wilson 1965)), and the center-to-limb dependence of spot-intensity contrast.

Interestingly, the time separation between consecutive transits is very close to the solar-rotation period (compare sine functions and TSI time-series in Figure 3.1a). A fluctuating behaviour of transit amplitude (i.e., decrease of the amplitude from first to second transit and increase of the amplitude from second to third transit) suggests that we are observing transits of different spots, even though the emergence of all three spots occurred at the same location on the solar surface. We confirm this by comparing simultaneous records of the MDI intensity maps and TSI [4]. Such a nesting of spot emergence (Brouwer and Zwaan 1990; Gaizauskas et al. 1994) affects the power spectrum of brightness variations exactly as the increase of the lifetime of the magnetic features. In particular, it makes light-curves more regular and helps to determine rotation period.

We perform a comparison between the GLS, ACF, PS, and GPS methods for the considered epoch of spot transits. In Figure 3.1b, we plot the results of the GLS analysis. A prominent peak is clearly observed at 27.2 days, which is very close to the *synodic*

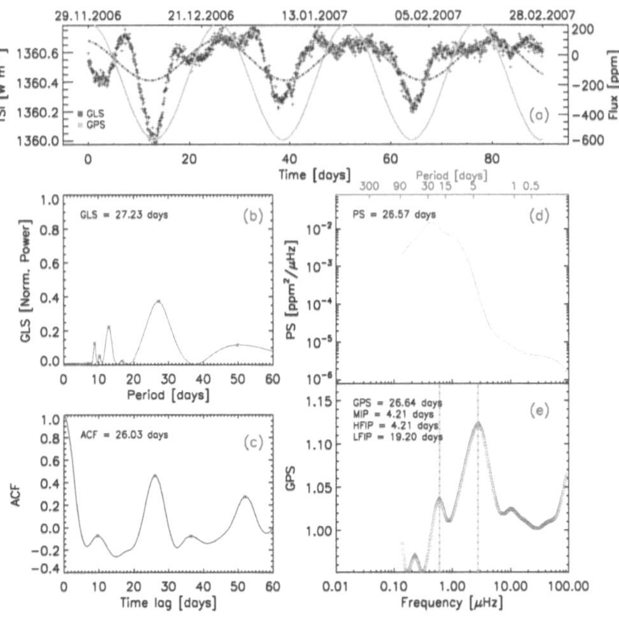

Figure 3.1: Example of the spot-dominated total solar irradiance (TSI) variability. panel (a) shows SORCE/TIM measurements from 28-Nov-2006 to 28-Feb-2007. Purple and orange curves indicate sine-wave functions with periods of 27.3 d (solar-rotation period deduced with the generalised Lomb Scargle periodogram method, GLS) and 26.6 d (solar-rotation period deduced with the gradient of power spectra method, GPS). Panels (b) and (c) show the corresponding GLS periodogram and autocorrelation function, ACF, respectively. Panel (d) shows the global wavelet power spectrum, PS, calculated with the 6th order Paul wavelet. Panel (e) shows the gradient of the power spectrum plotted in panel (d). Blue asterisk signs in panels (b) and (c) represent positions of the peaks in the GLS and ACF. Green dotted lines in panel (e) indicate the high- and low-frequency inflection points.

Carrington rotation period of 27.27 days. This peak has a power of 0.37, where GLS is normalised to unity. A second peak appears at 13.1 days with a normalised power of 0.23, which is a harmonic of the solar-rotation period. A large value of the normalised power in the rotation peak is not surprising: the corresponding light-curve has a clear periodicity, which results in a clear peak in the power spectrum. The purple dashed line in Figure 3.1a fits a sine wave with period determined from the GLS method.

Figure 3.1c shows the ACF analysis. Two clear peaks are visible at 26.0 and 52.0 days, having amplitudes of about 0.5 and 0.3, respectively. PS analysis is shown in Figure 3.1d. We use a Paul wavelet basis function with order m=6 to calculate the global wavelet power spectrum. PS show a pronounced peak around 26.6 days. One can also see a shoulder-like feature at about 13 days that is generated due to the ingress and egress of magnetic features over the visible solar disk. The contrast differences between faculae and spots transiting the limbs and the center will imprint a characteristic pattern in the light-curve, detected by the implementation of the GPS.

Figure 3.1e shows the GPS profile with two well-defined peaks that correspond to the inflection points in the power spectrum. The low-frequency inflection point is at 19.2 days and is associated with the transition from the rotation peak to high-frequency modulations forming a shoulder-like feature. The high-frequency inflection point is at 4.21 days and corresponds to the transition from the shoulder-like feature to the plateau at higher frequencies. For the considered period of brightness variations due to spots, the high-frequency point also corresponds to the maximum value of the GPS. In other words, the gradient is larger at the high-frequency point than at the low-frequency point.

In Sect. 3.3.5, we show that, in agreement with Paper I, the high-frequency inflection point is present in the PS even when the rotation peak is absent. Furthermore, the location of the high-frequency inflection point is stable independently of the presence or absence of the rotation peak.

Based on the location of the high-frequency inflection point at 4.21 days and using Eq. (3.1) with the calibration factor $\alpha_{\mathrm{Sun}} = 0.158$ and its 2σ uncertainty of 0.014 we compute the solar-rotation period: $P_{\mathrm{rot}} = 26.6^{+2.2}_{-2.6}$ days. We use the 26.6-day value for plotting the orange sine curve in Figure. 3.1a. This rotation period is in reasonably good agreement with the Carrington period given the relatively short length of the time-series.

In summary, all methods applied above allow reasonably accurate determinations of the rotation period for the considered period of pseudo-isolated transits of sunspots.

3.3.3   Brightness signature of facular feature transits

Here we perform a rotation period analysis of an interval of time between December 2007 and March 2008 (see Figure 3.2) when the TSI variability was dominated by three consecutive transits of a facular feature.

The simultaneous analysis of MDI intensitigram, magnetograms, and TSI records [56] shows that variability is brought about by a single large facular feature, which size is decreasing from transit to transit. We note the difference with the spot case described in Sect. 3.3.2 where the three consecutive transits were caused by the nesting effect. Such a behaviour is consistent with the fact that lifetimes of facular features are significantly larger than that of the spots.

The transit of a facular feature has a characteristic double-peak "M-like" profile, see Figure 3.2a. This "M-like" profile can be explained by the increase of facular contrast towards the limb. Such an increase partly compensates the foreshortening effect and, consequently, the maximum brightness occur when the facular feature is observed at an intermediate disk position.

Figure 3.2b shows a GLS analysis of the considered time interval. A prominent peak is seen at 26.4 days with a normalised power of 0.68. The purple dash line in Figure 3.2a fits to the light-curve with a sine wave with a corresponding period. Figure 3.2c shows the

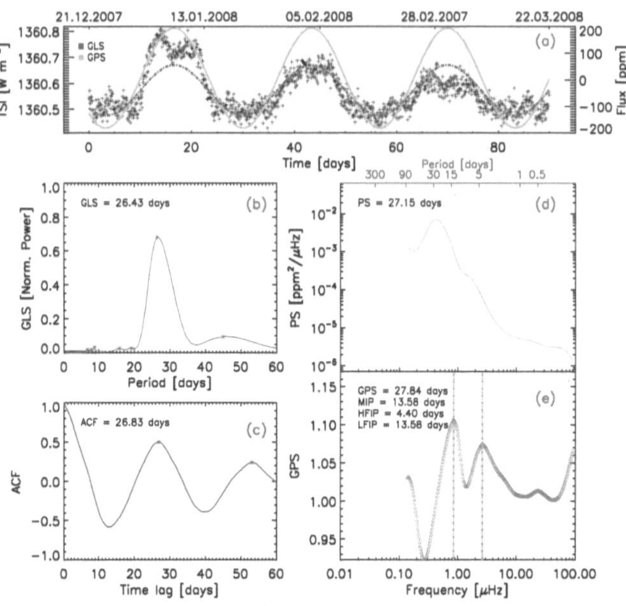

Figure 3.2: The same as Figure 3.1 but for the time interval of faculae-dominated TSI variability from 21-Dec-2007 to 22-Mar-2008. The purple dashed curve in panel (a) represents a sine wave function with a period of 26.4 d corresponding to the solar rotation period deduced with the GLS method. The orange curve in panel (a) is a sinusoidal function with a period of 27.8 d corresponding to the solar rotation period deduced with the GPS method.

corresponding ACF analysis. A clear series of peaks with time lag 26.8 days are observed.

The PS analysis shows a peak at 27.2 days (see Figure 3.2d). Similarly to the case of spot transits one can see a shoulder-like feature, but it is shifted towards higher frequencies in comparison to the spot case. Such a shift of the shoulder-like feature is explained by the M-like profile of a facular transit in the light-curve. It leads to the enhanced variability on timescales shorter than the half of the rotation period and consequently shifts the shoulder-like feature.

Figure 3.2e shows the GPS profile with two clearly visible peaks that corresponds to the inflection points of the PS. As in the case of the variability brought about by sunspots there are two inflection points, one is closer to the rotation period peak and another to the shoulder-like feature. We note that for the faculae-dominated case, in contrast to the spot-dominated case, the maximum inflection point (i.e., the inflection point with highest value of the GPS) corresponds to the low frequency inflection point.

The high-frequency inflection point is located at 4.40 days, which is very close to the location of the inflection point during the spot-dominated regime of the variability

(4.21 days, see Sect. 3.3.2). Applying the calibration factor α_{Sun} one can see that the value given by the GPS method for the solar rotation period is $P_{rot} = 27.8^{+2.3}_{-2.7}$ days. The orange line in Figure 3.2a fits a sinusoidal function with a period of 27.8 days as returned by the GPS analysis. The amplitude of the orange curve is given by the maximum amplitude of the wavelet.

The analysis performed in this Section and in Sect. 3.3.2 shows that when solar brightness variability is attributed to the periodic transits of spots and faculae all four methods (GLS, ACF, PS, and GPS) can accurately retrieve solar rotation period. However, the recurrent transits of one spot group or nested spots (as plotted in Figure 3.1a) are rare. Since faculae last significantly longer than spots, the recurrent transits of the same facular features are much more common than that of spots. However, faculae dominate the solar TSI light-curve on rotational timescales only at activity minimum (when no large spots are present, although on the solar cycle timescale, faculae play a dominant role).

Most of the time solar brightness variations are brought about by the combination of magnetic features coming with random phases. As we will show below this strongly affects the performance of the GLS, ACF, and PS methods but has a much weaker effect on the GPS method.

3.3.4   Analysis of the entire data-set

In Sect. 3.3.2 and 3.3.3 we considered relatively short intervals of time when solar variability is dominated by either spot or facular components. In this section we study whether the solar rotation period can be reliably retrieved in a more general case when contributions from faculae and spots are entangled. For this, we consider 21 years of TSI data from SoHO/VIRGO (see Figure 3.3) and 15 years of TSI SORCE/TIM data (see Figure 3.4) and apply GLS, ACF, PS, and GPS methods to the entire time-series.

We plot GLS periodograms for VIRGO and TIM data in panels (b) of Figs. 3.3 and 3.4, respectively. One can see that none of the two periodograms contain a sufficiently strong peak to provide a clear indication of periodicity. Instead they contain a series of peaks with similar and relatively small power: up to 0.003 for the VIRGO TSI and 0.007 for the TIM TSI. This is about two orders of magnitude lower than the normalised power obtained for pseudo-isolated spots and facular cases considered in Sects. 3.3.2 and 3.3.3. Consequently, the GLS method does not allow a definitive detection of the rotation period when the entire TIM and VIRGO time-series are considered. The ACF of the VIRGO and TIM TSI time-series are shown in panels (c) of Figs. 3.3 & 3.4.

Although we can appreciate small peaks at the expected location of the rotation period in both ACFs cases, the significance of the maxima with the corresponding time lag needed for the identification of the rotation period yields only a marginal detection. Panels (d) in Figures 3.3 and 3.4 display the PS analysis of the data. One can see that instead of the peak at the rotation period, one can observe a plateau region for both data-sets.

All in all the, GLS, and PS methods cannot detect the solar rotation period, and ACF give us a marginal detection when long sets of TSI data are considered. This is in line with the result of Shapiro et al. (2017) who showed that the superposition of facular and spot contributions to solar variability can significantly decrease the rotation signal.

The GPS profiles of VIRGO and TIM time-series are given in Panels (e) of Figures 3.3 and 3.4. One can see that both profiles display conspicuous high frequency inflection

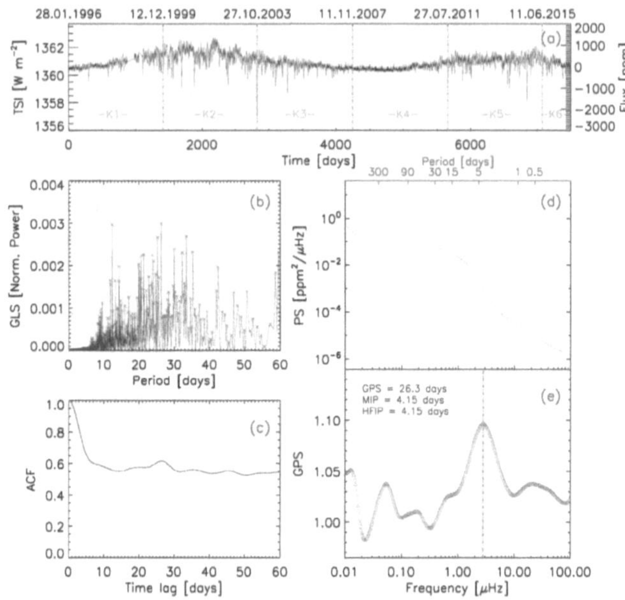

Figure 3.3: The same as Figure 3.1 but for 21 years [1996.01.28–2017.05.23] of TSI data from SoHO/VIRGO. In total 7787 days are considered. This corresponds to about 186889 data-points at an hourly cadence. The GLS, ACF, and PS methods do not show a clear signal of the rotation period. GPS shows a prominent peak at 4.15 days, resulting in a solar rotation period value of about 26.3 days (see text for details). Orange vertical lines and marks in panel (a) represent a splitting of the VIRGO data-set into Kepler-like time-span K_n, each one representing the full lifetimes of the *Kepler* mission, around 4 years. (see Sect. 3.3.5).

points. They are located at 4.15 and 4.12 days for VIRGO and TIM data, respectively. Consequently, we retrieve a rotation period for the Sun: $P_{\rm rot} = 26.3^{+2.1}_{-2.5}$ days, and $P_{\rm rot} = 26.1^{+2.1}_{-2.5}$ days, for the VIRGO and TIM data, respectively. These $P_{\rm rot}$ values agree within the error bars with the solar *synodic* Carrington rotation period of 27.27 days, as well with the solar equatorial *synodic* rotation period for a fixed feature of 26.24 days. Consequently, the GPS method allows a proper determination of solar rotation period over timescales in comparison with other traditional approaches.

3.3.5   The solar variability in 90-day quarters

Here we consider the exemplary case of the Sun as it would be observed over the same time-span as the *Kepler* stellar data-set. The *Kepler* telescope reoriented itself every 90 days, thus introducing discontinuities into the light-curves. To mimic the observational

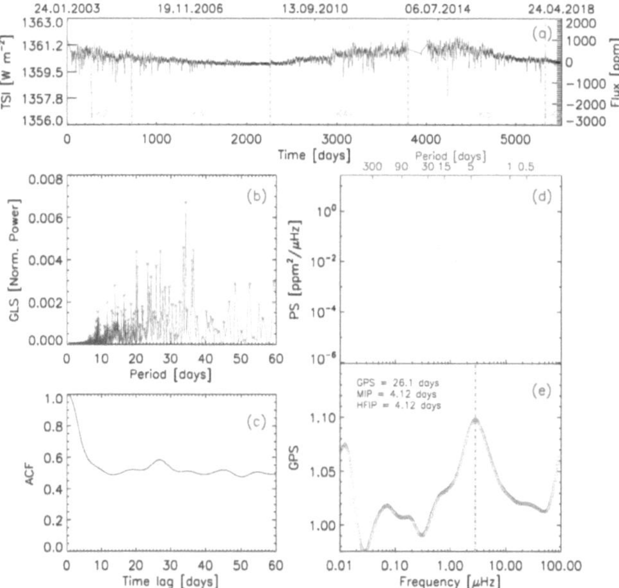

Figure 3.4: The same as Figure 3.1 but for 15 years [2003.01.24–2018.04.24] of TSI data from SORCE/TIM. In total 5508 days are considered. This corresponds to about 82632 data-points at a cadence of 1.6 hours. Orange lines represent the same Kepler-like time-span K_n than shown in Figure 3.3. The GLS, ACF, and PS methods do not show a definitive detection signal of the rotation period. GPS method shows a prominent peak at 4.12 days corresponding to the solar rotation period value of 26.1 days.

routine of the *Kepler* telescope, we segment the entire VIRGO and TIM data-sets into 86 and 60 quarters of 90-day duration, with a cadence of 1.0 and 1.6 hours, respectively. The cadence of solar observations are close to regular long cadence *Kepler* observations of 29.4 min. In this section we analyse the stability of the rotation signal from quarter to quarter, using the four methods described above.

The inflection points described by the GPS method in all quarters are shown in Figure 3.5 for the VIRGO (top panel) and TIM (bottom panel) data. We note that quarters 10, 12, 13 and 52 in the VIRGO data-set are affected by the lack of data due to spacecraft and instrument failures. TIM data quarters number 1, 43, 44, and 45 contain long gaps, some of them larger than 5 days. For the three recent quarters this is because of the failing SORCE battery.

We note that power spectra of some quarters have more than one inflection point with considerable amplitude (like in the time intervals shown in Figs. 3.1 and 3.2). We analyse the two highest inflection points in the GPS and describe these in terms of its frequency location. The inflection point towards the highest frequency, HFIP, is represented by blue

3 Inflection point in the power spectrum of stellar brightness variations. II. The Sun

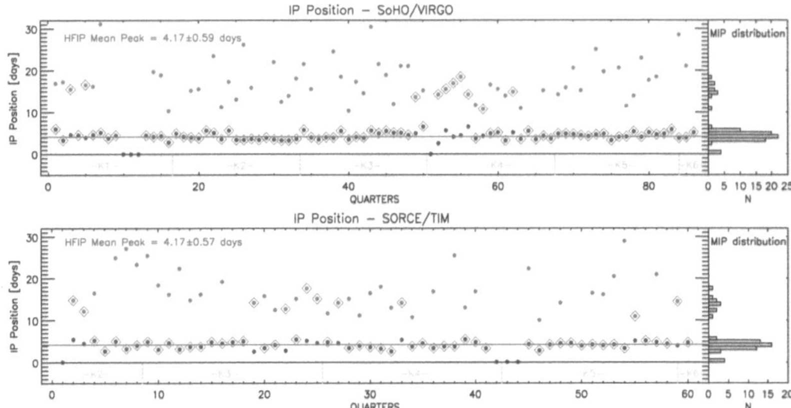

Figure 3.5: Top-Left panel: Positions of the inflection points for the 86 (90-day) quarters of the VIRGO data. Red dots represent the low frequency inflection points (LFIP), blue dots represent the high frequency inflection points (HFIP), black diamonds indicate inflection points with maximum GPS value (see text). Top-Right panel: Distribution of the maximum inflection point positions. Bottom panels: the same as top panels but for 60 quarters of 90-days using the TIM data. Orange lines in left panels indicate splitting of the VIRGO and TIM data-sets into Kepler-like time-span K_n as in Figure 3.3.

dots, in Figure 3.5. The inflection point towards lower frequencies, LFIP, is represented by red dots. In Figure 3.5 for each quarter open black diamonds surround the inflection point corresponding to the maximum amplitude of the gradient of the power spectra, MIP. In most of the cases these maximum peaks are simultaneously the high frequency inflection points (as in the case of the variability brought about by spots, see Figure 3.1), marked with open black diamonds over-plotted over the blue dots. Similarly, there are several quarters when the maximum value of the gradient corresponds to the low frequency inflection point (like in the case of the variability brought about by faculae, as we explain before, see Figure 3.2). These quarters are associated with low solar activity and with TSI variability being dominated by faculae. Interestingly, the high frequency inflection points are still present in such quarters (even though they no longer correspond to the maximum amplitude of the gradient). In Figure 3.5 we indicate when the maximum GPS amplitude corresponds to the low frequency inflection points with open black diamonds over-plotted over red dots.

Figure 3.5 demonstrates that positions of the inflection points are stable and the mean value of positions over all considered quarters are basically the same for the VIRGO and TIM data. We construe this as the prove that the GPS method works for the Sun.

Having shown that the position of the inflection point is stable for the Sun, we could then use the Sun to calibrate the GPS method and apply α_{Sun} value to other Sun-like stars, independently of the simulations presented in Paper I. At the same time, it is reassuring to see that the theoretical α_{Sun} value found in Paper I leads to a reasonable value of the solar rotation period. Indeed, after applying the α_{Sun} calibration factor from Paper I to the position of the inflection point at 4.17 ± 0.59 d for VIRGO data and 4.17 ± 0.59 d for TIM data (where 0.59 d and 0.57 d corresponds to 2σ of the observed distribution of values for all quarters) we obtain 26.4 ± 3.7 d and 26.4 ± 3.6 d, respectively.

We note that the uncertainty of the rotation period is calculated here in a different way than in Sects. 3.3.2–3.3.4, where it was defined via the uncertainty of the theoretical α_{Sun} value (σ_α). This σ_α uncertainty is brought about by the dependence of the inflection point position on the specific realisation of emergence of magnetic features. In contrast, in this section we calculate the rotation period and its uncertainty as:

$$P_{\mathrm{rot}} \pm \delta P = \frac{(\overline{HFIP} \pm 2\sigma_{HFIP})}{\alpha_{Sun}}, \tag{3.2}$$

where the σ_{HFIP} value is the standard deviation of the observed distribution of inflection point positions (blue points in Fig. 3.5). As well as σ_α, σ_{HFIP} accounts for the uncertainty due to the randomness in emergence of magnetic features (so that it does not make sense to account for σ_α in Eq. 3.2). In addition, σ_{HFIP} also accounts for the noise in the TSI data. Hence, we utilise here Eq. (3.2) and σ_{HFIP} for estimating the uncertainty of the rotation period with the GPS method.

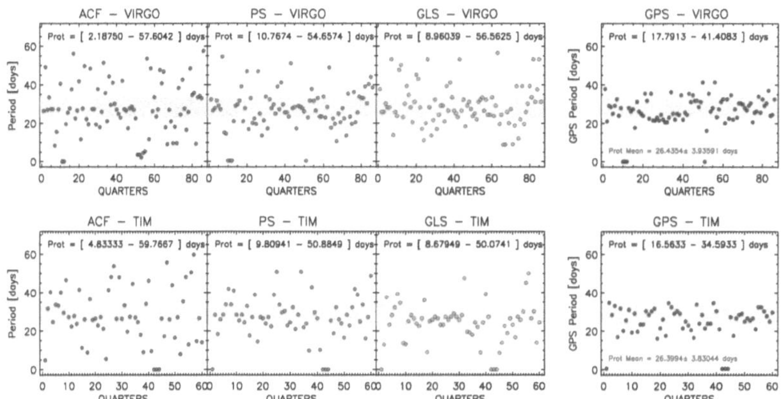

Figure 3.6: Values of the rotation periods per 90-day quarters returned by the ACF, PS, GLS, and GPS methods. The analysis is performed for the VIRGO (top panels) and TIM (bottom panels) data. Pale orange shaded areas cover the period range of [23–34] days (see Sect. 3.3.5 for details). Information about the ranges of rotation period values obtained by each method for different instruments is shown near the top of each panel.

Figure 3.6 compares the performance of the ACF, GLS, PS, and GPS methods. It shows one value of the rotation period per 90-day quarter determined with the four methods for 21 years of TSI by VIRGO (top panels) and 15 years of TSI by TIM (bottom panels). In addition, the pale orange bar denotes a range between 23 and 34 days, which we take to be the success range for determining the solar rotation period (which we expect would lie in the range 27–30 d within 4 d error bars). One can see that the distribution of retrieved periods is similar for data from both instruments.

3   Inflection point in the power spectrum of stellar brightness variations. II. The Sun

Met/D-Set	TIM_{Spot}	TIM_{Fac}	$VIRGO_{21Y}$	TIM_{15Y}	$VIRGO_Q$	TIM_Q
ACF [d]	26.0	26.8	–	–	2.2-57.0	4.8-59.8
S [%]	–	–	–	–	37.2	36.1
GLS [d]	27.2	26.4	–	–	8.9-56.6	8.6-50.1
S [%]	–	–	–	–	55.0	47.0
PS [d]	26.5	27.1	–	–	10.8-54.6.	9.8-50
S [%]	–	–	–	–	51.2	54.1
GPS [d]	$26.6^{+2.2}_{-2.6}$	$27.8^{+2.3}_{-2.7}$	$26.3^{+2.1}_{-2.5}$	$26.1^{+2.1}_{-2.5}$	7.8-41.4	16.6-34.6
S [%]	–	–	–	–	88.4	82.1
HFIP [d]	4.21	4.40	4.15	4.12	4.17±0.59	4.17±0.57
LFIP [d]	19.20	13.58	–	–	–	–

Table 3.1: Compilation of the rotation period analysis for the four different methods implemented (Met) (autocorrelation functions (ACF), generalised Lomb-Scargle periodogram (GLS), wavelet power spectra (PS), and gradient of the power spectra (GPS)) and its success percentage for the different data-sets used in this work (TIM SPOT, for the pseudo-isolated spot transit, TIM FAC, for the pseudo facular region transit, VIRGO [21 Y] for the entire VIRGO data-set, TIM [15 Y] for the entire TIM data-set, VIRGO [Q] for the VIRGO data-set analysis per quarter, TIM [Q], for the TIM data-set analysis per quarter).

The values obtained by ACF (Maroon dots) range between 2.18 days to 57.6 days for VIRGO data (see left panels of Figure 3.6). There are quarters where ACF can accurately detect solar rotation period, but there are many other quarters where the method fails. Overall the rotation values retrieved by the ACF lie between 23 and 34 days for 32 out of the 86 VIRGO quarters, i.e., the ACF method has a success rate of 37.2 % when applied to the VIRGO data. For the TIM data-set the ACF obtained values are in between of 4.8 to 59 days. The success rate is 36.1 %, as shown in Table 3.1.

Second from the left panels in Figure 3.6 show the performance of the PS method. The retrieved rotation periods are in between 10.7 and 54.7 days for the VIRGO data and in between 9.8 days and 50.8 days for the TIM data. The success rates for period determination are 51.2 % and 54.1 % for VIRGO and TIM, respectively.

Using GLS we are able to retrieve closer solar periodicities per quarter as it is shown in the right panels in Figure 3.6. GLS successfully retrieves the solar rotation period for 55 % of the quarters using VIRGO data and 47 % for TIM data. We notice lower scatter for the TIM data-set, with the returned rotation periods lying in a range between 8.6 and 50.0 days. For VIRGO data we obtain rotation period between a range of 8.9 to 56.7 days.

In the right panels of Figure 3.6 and Table 3.1 we show values retrieved with the GPS method. One can see that the GPS method results in less scatter for the retrieved values of the rotation period, finding rotation periods in the range [17.8-41.4] days for VIRGO data and [16.6-34.6] days for the TIM data-set. The GPS method achieves success rates of 88.4 % and 82.1 % for the VIRGO and TIM data, respectively. The regularity of the signal allow us to analyse the distribution of the high frequency inflection point and its behaviour over time.

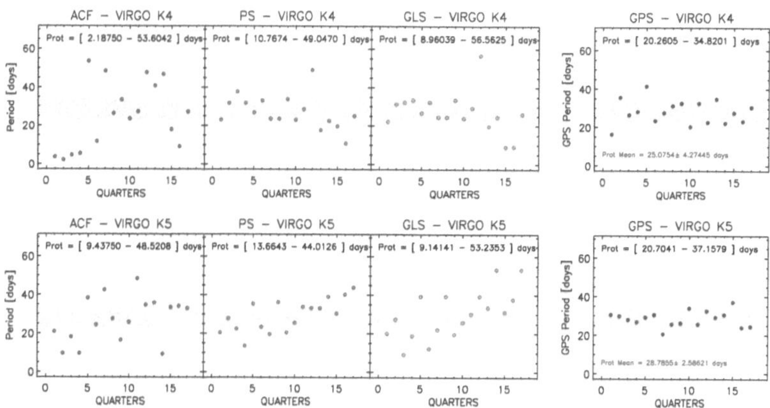

Figure 3.7: The values of the solar rotation period per 90-day quarters of the VIRGO data returned by the ACF, PS, GLS, and GPS methods (from left to right panels). Shown are the K_4 [2007:09:12 - 2011:07:27] and K_5 [2011:07:28 - 2015:06:11] Kepler-like time-span, corresponding to low and high levels of solar activity, respectively. Pale orange colour areas indicate the period range of [23–34] days.

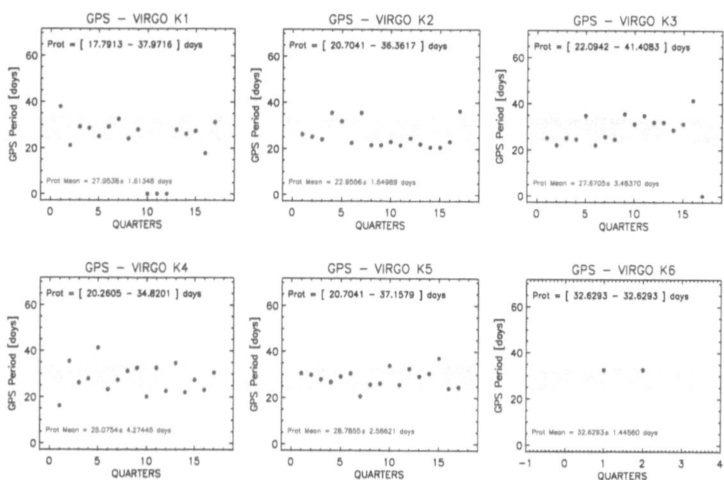

Figure 3.8: The same as Figure 3.7 but for all Kepler-like ($K_1 - K_6$) time-span of the VIRGO data. Only GPS values are shown. Blue dots represent the estimation of the rotation period per quarter obtained by GPS.

Regular stellar photometric observations are normally performed during unknown stellar activity stages. To characterise the detectability of the rotation period for the different activity time-spans, we test the performance of the GPS method in comparison with the ACF, GLS and PS methods for periods of relatively low and high solar activity. For this we use just VIRGO data, since it covers both periods of very high and low solar activity.

To mimic *Kepler* observations we split the entire period of VIRGO observations into five segments $K_1 - K_5$ (the length of the segments roughly corresponds to the total duration of the 4 years of *Kepler* observations) and the remaining 712-day segment K_6 (see vertical dashed orange lines in Figures 3.3, 3.4 and, 3.5). We then subdivided each K_n segment in 17 quarters of 90 days each, and analyse them separately.

The performance of all four methods is compared in Figure 3.7 for segment K_4 (corresponding to a period of low solar activity) and K_5 (high solar activity). We observe that for the low period of activity, K_4, the values obtained per quarter for PS, GLS show less scatter than for the values shown during high levels of solar activity in K_5 segment. ACF values show similar scattered rotation values for both, high and low levels of solar activity. The GPS method recover rotation period values closer to the solar rotation period range for both K_4 and K_5 segments.

Figure 3.8 shows the rotation period detected with the GPS method per quarter for K_1 to K_6 segments. While there is some scatter in the values of the rotation periods deduced from the analysis of segments K_1 to K_6 (in particular, values obtained in segment K_2 are 4 days lower than *sidereal* Carrington rotation period), Figure 3.8 indicates that the solar rotation period can be successfully retrieved by GPS for all K_n analysed segments.

3.3.6   The impact of white noise in the inflection point position

In Subsection. 3.3.5 we have processed solar TSI data to represent the Sun as it would be observed with the time-span of *Kepler* observations. We have considered two TSI data-sets, one obtained by SoHO/VIRGO, another by TIM/SORCE. While the noise level in these two data-sets is rather different (Kopp 2016), the positions of the inflection points are basically independent off the data-set (see, Figs. 3.5 and 3.6). This implies that our analysis is only weakly affected by the noise in TIM and VIRGO data. At the same time the noise level in *Kepler* data normally is significantly higher than those in the solar data.

Solar and stellar light-curves are recorded in a different way so that the noise sources are also substantially different. TSI is measured using radiometers while stellar photometric measurements are performed using Charged Coupled Devices (CCD). Photon detection by a CCD is a statistical process associated with several sources of noise, which can be generally approximated by Gaussian white noise.

To assess the impact of noise on the position of the inflection point we artificially added white noise to the VIRGO TSI data. The amplitude of the noise was chosen following the expected dependence (see Van Cleve and Caldwell 2016) of the noise value per specific *Kepler* magnitude Kp, that is the measured source intensity observed through the *Kepler* bandpass.

Figure 3.9 shows the position of the high frequency inflection point as a function of the white noise level. The colours of the dots represent the *Kepler* magnitude Kp. For each Kp-value (and, corresponding, amplitude of the white noise) we calculate five realisations of the noise, add it to the entire VIRGO data-set shown in Figure 3.3, and calculate the position of the inflection point as in Sect. 3.3.4. One can see that the inflection point shifts

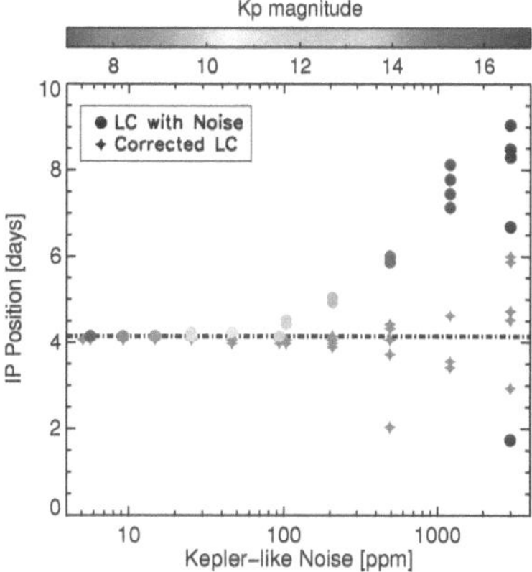

Figure 3.9: Position of the high frequency inflection point calculated for the entire VIRGO data-set (as in Figure 3.3) as a function of Kepler-like white noise added to the original VIRGO data. The expected level of white noise is a function of the stellar *Kepler* magnitude, Kp (see main text for more information). The colour of the dots (see colour panel in the top of the figure) indicates Kp magnitude corresponding to the expected level of the white noise. Grey star symbols represent the position of the inflection point after the noise correction (see text for details). There are five different realisations of noise per each value of the Kp magnitude. When the level of noise in the LC is lower than 100 ppm, around Kp=12, the values of IP values are overlapped and appear as a single point in the plot.

to lower frequencies when the noise level is increased.

We have tested a simple method for mitigating such a shift. Namely, we utilized the fact that the power spectrum flattens at high frequencies. The power in the flattened part represents the superposition of white noise and granulation (Shapiro et al. 2017). We have calculated the mean power between periods of 1 hour and 1 day and subtracted this single value from the entire power spectrum. Then, we recalculated the position of the inflection point. These corrected positions of the inflection points are represented by grey filled star symbols in Figure 3.9. One can see that our method is reasonably effective until the noise level reaches about 300-400 ppm, which corresponds to about a Kp magnitude of 14 for 1-hour cadence light-curve.

Beyond a level of introduced noise of 500 ppm the error in the estimation of the real high frequency inflection point location starts to become considerable, even though the location of the high frequency inflection point still gets more accurate after the correcting noise procedure, see grey star symbols in Fig. 3.9. For example, for a star with the level of noise expected for a Kp magnitude of 15, the mean scatter in the high frequency inflection point value corresponds to 0.83 days, which yields to a 5.25 days deviation in the rotation period value.

3.4 GPS and skewness relation

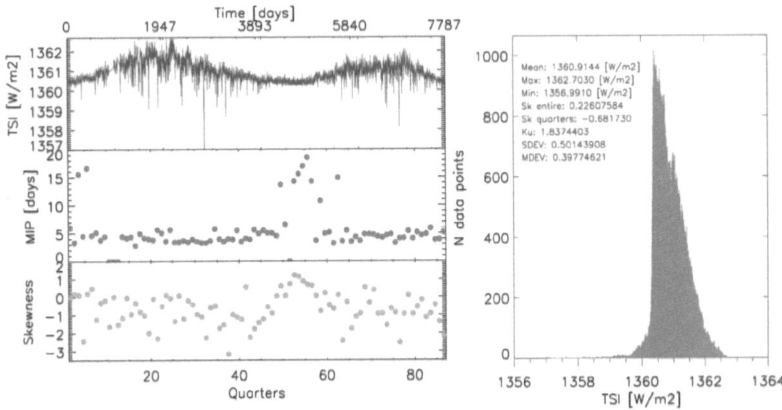

Figure 3.10: Top panel of left image: 21-years of TSI data gathered by VIRGO. Middle panel of left image: Positions of the maximum inflection point (MIP) per quarter. Left-Bottom: Skewness values per quarter. Right: Distribution of the TSI values shown in the top-left panel. We list mean, maximum, and minimum TSI values, skewness for the entire data-set (Sk over the whole time-series), the mean of individual skewness values calculated per quarter (Sk quarters, see text for more details), kurtis (Ku), standard deviation (SDEV), and mean deviation (MDEV).

Distinguishing between facular- and spot-dominated regimes of brightness variability is important for understanding the structure of the stellar magnetic field and for identifying biases in determination of stellar rotation periods.

While solar rotation variability is predominantly spot-dominated, there are also periods of facular domination (see Sect. 3.3.3). In Sects. 3.3.2 and 3.3.3 we demonstrated that the GPS spectrum has a different profile depending on whether variability is facular- or spot-dominated. In this section we show that in the solar case these two regimes can also be distinguished based on the skewness of the distribution of TSI values. This suggests that skewness can be a good indicator of the variability regime for low-activity stars like the Sun.

For a data-set of TSI values the skewness can give us valuable information about the distribution of maximum and minimum values. In other words when we have a decrease of the intensity due spots the distribution will be skewed preferentially towards the left side (i.e., to the lower values) of the maximum peak of the distribution. When an increase of intensity is registered in the light-curve due the presence of brighter facular regions the skewness will shift to the right side (i.e., to the higher values) of the distribution. The skewness of a distribution can tell us about its degree of symmetry.

In order to analyse the relation between skewness and the regime of solar variability we calculate skewness of the TSI values in each of the 90-day quarters introduced in

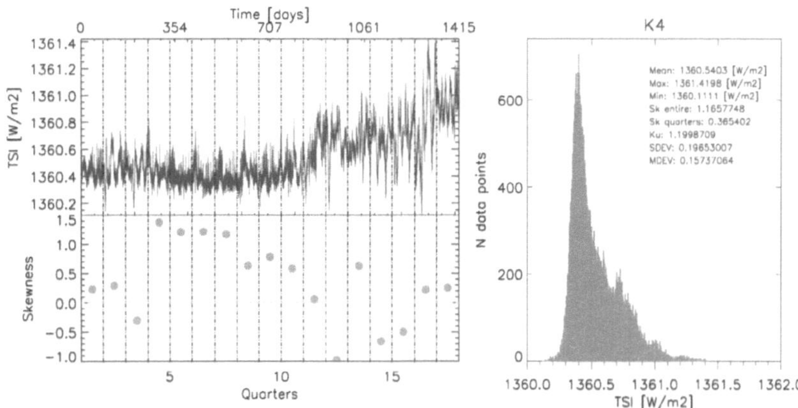

Figure 3.11: Skewness analysis for low-activity Kepler-like time-span K_4. Upper left panel: VIRGO TSI time-series during the K_4 season, which consists of 1415 days emulating 17 Kepler quarters. Lower left panel: Skewness values per quarter. Right: the same as right panel in Figure 3.10 but showing only TSI values in K_4 season.

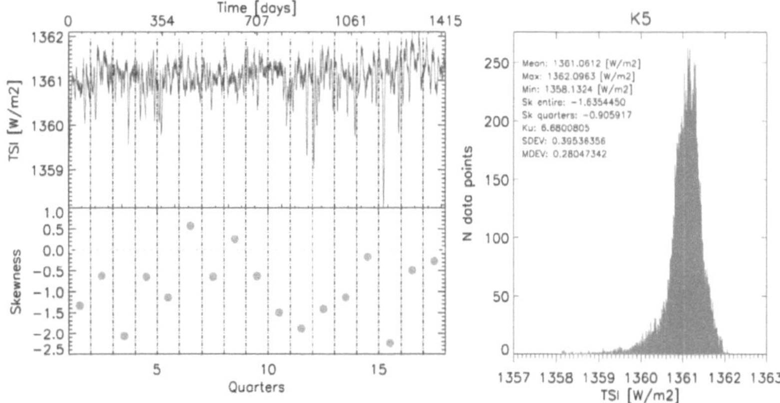

Figure 3.12: The same as Figure 3.11 but for high-activity Kepler-like time-span K_5.

Subsection. 3.3.5. In the upper left panel of Figure 3.10 we show the 21-year span of VIRGO TSI data. In the middle left panel we illustrate the location of the maximum inflection point (see Sect. 3.3.4) per quarter. As discussed in Sects. 3.3.2 and 3.3.3, the maximum inflection point (MIP) corresponds to the low frequency inflection point for faculae-dominated regimes and high frequency inflection point for spot-dominated regimes. The bottom left panel shows the skewness values for all 90-day quarters. One can see that periods of low solar activity, when TSI variability is mainly brought about by faculae (see, e.g., discussion in Shapiro et al. 2016), simultaneously correspond to positive skewness and maximum GPS value reached at the low frequency inflection point. The observed scatter in skewness values is higher scatter than the scatter in the MIP by GPS. This implies that the inflection points analysis provides a better indication of when the LC is mainly drawn by spot or facular components.

In the right panel of Figure 3.10 we show the distribution of TSI values for the entire VIRGO data. The skewness for the entire data-set, $Sk_E = 0.23$, is positive. This is because the skewness value of the entire data-set is affected by the TSI variability on the timescale of the 11-year cycle, which is faculae-dominated. To remove the contribution from the 11-year variability we can also calculate skewness by averaging all individual values per 90-day quarters. Since rotation TSI variability is mainly spot-dominated we then get a negative value of $Sk_Q = -0.69$.

Figures 3.11 and 3.12 show the skewness analysis for minimum and maximum activity segments, K_4 and K_5, respectively. The comparison between the light-curve segmented in quarters and its respective skewness values are in agreement with conclusions drawn analysing the entire VIRGO data-set. In particular, one can see that quarters with prominent positive excursions of brightness caused by faculae correspond to positive skewness, while quarters with negative excursions caused by spot correspond to negative values of skewness. Clearly, also the skewness of brightness distribution in the entire K_4 segment (corresponding to the minimum of solar activity) is positive, while skewness values for the K_5 segment with higher value of activity is negative. We will extend the combined, skewness and GPS, analysis to stars observed by *Kepler* and TESS in the forthcoming publications.

3.5   Discussion & Summary

The determination of rotation periods of stars with activity levels similar to that of our Sun is a challenging task, even when using high quality data from space-borne photometric missions. In Shapiro et al. (2020) we have proposed the GPS method specifically aimed at the determination of periods in old inactive stars, like our Sun. The main idea of the method is to calculate the gradient of the power spectrum of stellar brightness variations and identify the inflection point, i.e., the point where concavity of the power spectrum changes its sign. The stellar rotation period can then be determined by applying a scaling coefficient to the position of the inflection point.

We have applied the GPS method to the available measured records of solar brightness (specifically the total solar irradiance) and compared its performance to that of other methods routinely utilized for the determination of stellar rotation periods.

There are time intervals when solar light-curve has a regular pattern, the GPS and other methods, return correct value of solar rotation period. These intervals correspond to low values of solar activity when variability is either brought about by long-living faculae or nested sunspots. However, most of the time, solar brightness variations are attributed to superposition of simultaneous contributions from several bright and dark magnetic features with random phases. We have shown that this leads to a failure of other methods to identify a clear signal of the rotation period. At the same time, the GPS method still allows an accurate determination of the rotation period of the Sun independently of its activity level and the number of features contributing to brightness variability and of the ratio of facular to sunspot area.

In particular, we have shown that GPS method returns accurate values of solar rotation period for most of the time-span of SoHO/VIRGO and SORCE/TIM measurements, with exception of several intervals affected by the absence of data. We found that when the entire 21-year VIRGO and 15-year TIM data-sets are split in Kepler-like 90-day quarters and inflection points are calculated for each of the quarters, the maximum of the distribution of the inflection point positions peaks at 4.17 ± 0.59 days for VIRGO data-set and 4.17 ± 0.57 days for TIM data-set (see Figure 3.5). This results in a determination of the solar rotation period of 26.4 ± 3.7 days and 26.4 ± 3.6 days for VIRGO and TIM data-sets respectively. In a series of typical Kepler-like observations of the Sun, the GPS method can correctly determine the rotation period in more than 80 % of the cases while this value is about 50 % for GLS and below 40 % for ACF.

Typically solar variability on timescales up to a few months is spot-dominated. However, there are also time intervals when it is faculae-dominated (see, e.g., Figure 3.2). We have shown that these regimes can be distinguished from the GPS profile thanks to substantially different centre-to-limb variations of facular and spot contrasts. Furthermore, the two regimes can be separated by analysing the comparison between the inflection point location from GPS and the skewness of light-curves: the bright faculae lead to positively skewed light-curves and a stronger signal at the low frequency inflection point, while dark spots lead to negatively skewed light-curves and a dominant signal at the low frequency inflection point. However, the skewness values in Figure 3.10 show higher scatter than the IP by GPS. This implies that the IP by GPS provide a better indication of when the LC is mainly drawn by spot or facular components.

We construe the success of the GPS method in the solar case as an indication that it can be applied to reliably determine rotation periods in low-activity stars like the Sun, where other methods generally fail. Furthermore, our analysis demonstrates that photometric records alone can be used to identify the regime of stellar variability, i.e., whether it is dominated by the effects of spots or of faculae. In subsequent papers we will apply GPS method to determine rotation periods and regimes of the variability of *Kepler* and long term follow up of TESS stars.

4 Inflection point in the power spectrum of stellar brightness variations III: Facular versus spot dominance on stars with known rotation periods

4 Inflection point in the power spectrum of stellar brightness variations III: Facular versus spot dominance on stars with known rotation periods

4.1 Introduction of chapter 4

Rotation periods in cool main-sequence stars can be traced by observing the brightness modulation caused by the presence of active regions on stellar surfaces. Those active regions are generated by the emergence of strong localised magnetic fields approximately described by flux tubes (see e.g. Solanki 1993). Large flux tubes form dark spots, while ensembles of smaller flux tubes form bright faculae (see, e.g. Solanki et al. 2006, for a detailed review of the solar case). The active regions usually consist of a sunspot group surrounded by faculae. The transits of such active regions over the visible disk as the star rotates would cause brightness variability. Consequently, the stellar light curves (LCs) contain information about both the rotation periods and the properties of active regions. However, retrieving this information from the LCs often proves a daunting task (see e.g. Basri 2018).

The *Kepler* mission (Borucki et al. 2010) has provided the community with records of photometric observations with unprecedented precision and cadence. The *Kepler* LCs have been widely used to determine stellar rotation periods (e.g. Walkowicz and Basri 2013; Reinhold and Gizon 2015; Nielsen et al. 2013; García et al. 2014; McQuillan et al. 2014; Buzasi et al. 2016b; Angus et al. 2018; Santos et al. 2019). Despite considerable success in determining the rotation periods of many fast-rotating and active stars (see, e.g. McQuillan et al. 2014, who published rotation periods of about 34030 stars identified as being located on the main sequence) there is a lack information on periods of slowly rotating stars, that is, stars with near-solar and longer rotation periods. For example, the rotational period of the Sun may not be detectable during intermediate and high levels of solar activity (see Lanza and Shkolnik 2014; Aigrain et al. 2015).

The difficulties in detecting periods of slowly rotating stars might be an important contribution to the explanation of lower-than-expected numbers of G-type stars with near-solar rotation periods (van Saders et al. 2019). The difficulty in obtaining a reliable measurement of the rotation periods of stars with variability patterns similar to that of the Sun can also affect solar-stellar comparison studies (see e.g. Witzke et al. 2020; Reinhold et al. 2020b).

Within this context, we have developed a method aimed at determining rotation periods of low-activity stars like the Sun. In Shapiro et al. (2020) (hereinafter, Paper I), we found that the power spectra of brightness variations of such stars are strongly affected by the evolution of active regions. In particular, the rotation peak can be significantly weakened or it may even disappear from the power spectrum if the lifetimes of starspots are too short. Furthermore, the delicate balance between spot and facular contributions to the variability might lead to the appearance of spurious peaks, which do not correspond to the rotation period but could be easily mistaken for one (see also Shapiro et al. 2017).

In Paper I, we showed that the high-frequency tail of the power spectrum is much less affected by the evolution of magnetic features than frequencies near the rotation period. Consequently, we proposed to use information in the high-frequency tail for the determination of stellar rotation periods. In particular, we suggested that the period, P_{HFIP} , corresponding to the maximum of the gradient of the power spectrum (GPS) (i.e. to the inflection point) in the high-frequency tail could be used to identify the stellar rotation period, P_{rot}, via the simple scaling relation:

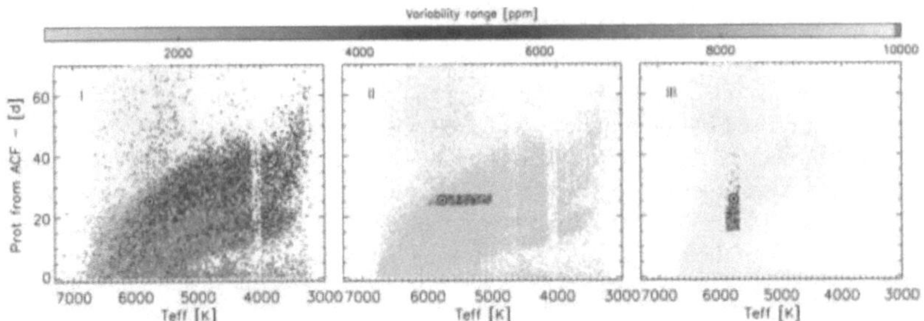

Figure 4.1: Panel I: temperature-rotation diagram for a sample of 34030 stars (coloured circles indicating the variability range) with rotation periods determined by McQuillan et al. (2014) and 55501 stars where they found a period but deemed it to be not significant (bisque dots, See panel-III for better visualisation). Panel II sample of 34030 stars with rotation periods determined coloured in grey and 55501 stars with not significant rotation period determination in bisque colour. For Panels II and III only stars from sample A (panel II) and sample B (panel III) are shown in colour, see Table 4.1 for the properties of samples A and B. Panels II illustrates the stellar sample A, selected by near solar rotation period and temperatures from 5000 K to 6000 K. Panel III, illustrates stellar sample B, that contains stars with near solar effective temperature and a broad range in rotation periods. The Sun is represented by the solar symbol ⊙.

Table 4.1: Stellar parameters for samples A and B.

Sample	N	$T_{\mathrm{eff}}^{(1)}$ [K]	log $g^{(1)}$	[Fe/H]$^{(1)}$	Var$^{(2)}$ range [ppm]	$P_{\mathrm{rot}}^{(2)}$ [d]
A	686	5000-6000	4.20-4.69	-1.46-0.56	211-39748	24.0-27.4
B	361	5700-5900	4.21-4.60	-1.08-0.44	211-17530	15.0-39.8
Sun	1	5778	4.44	0.0	300-1500	27.27 (Sy)
						25.38 (Sid)

Note. Stellar parameters for stellar samples A and B. 1) Effective temperature (T_{eff}), surface gravity (log g), and metallicity ([Fe/H]) values are taken from Huber et al. (2014). 2) Variability range (Var) and rotation periods (P_{rot}) are taken from McQuillan et al. (2014). We take the solar synodic (Sy) and sidereal (Sid) Carrington rotation period values as reference.

$$P_{\mathrm{rot}} = P_{\mathrm{HFIP}}/\alpha. \tag{4.1}$$

Here, α is a calibration factor which is independent of the evolution of active regions. It shows only a very weak dependence on the stellar inclination. For example, the inclination dependence can be neglected for inclinations of 45° and greater; see Fig. 9 from Paper I. Statistically, this corresponds to roughly 70% of stars.

The model developed in Paper I indicated that the value of α shows a moderate dependence on the ratio between facular and spot areas of the individual active regions at the moment of emergence, $S_{\text{fac}}/S_{\text{spot}}$. This ratio was assumed to be the same for all active regions (see a detailed discussion in Paper I). The dependence of the inflection point position on the facular-to-spot area ratio leads to a certain degree of uncertainty (up to 25%) when determining stellar rotation periods since the value of $S_{\text{fac}}/S_{\text{spot}}$ for a given star is a priori unknown. At the same time, it allows us to retrieve valuable information about facular versus spot-dominated regimes with regard to the variability of stars where rotation periods can be determined using other methods (see Amazo-Gómez et al. 2020b)hereafter, Paper II).

A first test of the gradient of the power spectrum method (hereafter, GPS) was performed in Paper II, where we applied it to solar brightness variations. We showed that in contrast to other methods, GPS allows for an accurate determination of the solar rotation period at all levels of solar activity. Additionally, we analysed time intervals where solar variability was spot-dominated and when it was faculae-dominated. We showed that these regimes can be distinguished in the GPS profile due to the substantially different center-to-limb variations of faculae and spots.

In this study, we apply the GPS method to stars with determined rotation periods from *Kepler* photometry. The goal here is twofold: firstly, we test the GPS method further before applying it to stars with unknown rotation periods; secondly, we investigate whether the α factor and, consequently, the facular or spot composition of stellar active regions, is dependent on the rotation period. In Section 4.2, we describe the stellar sample we used. In Section 4.3, we present the main results. Our conclusions are summarised in Section 4.4.

4.2   Stellar sample selection

In this study, we consider stars in the field of view (FOV) of the *Kepler* telescope for which McQuillan et al. (2014) managed to determine rotation periods using the auto-correlation function (ACF). To ensure that the main source of the variability for the selected stars is magnetic activity, we only selected stars on the main-sequence, using T_{eff} and log g values from the Huber et al. (2014) catalogue to exclude giants (see Table 4.1). We note that Huber et al. (2014) calibrated effective temperatures to the infrared flux temperature scale. This resulted in an approximately 200 K offset from the original Kepler Input Catalogue (KIC) (Pinsonneault et al. 2012). We also precluded stars flagged in the KIC as giant (GS), eclipsing binary (EB), or host stars with planetary transits confirmed (PTC), with planet candidates (PC), and false-positive planets (FP).

We selected two sets of stars with near-solar parameters. The selection criteria for both samples (A and B) are illustrated in Fig. 4.1 and given in Table 4.1. Figure 4.1 and 4.2 show the variability ranges of the set of selected stars. Kepler observatory provided four years of photometric information, from 2009 to 2013, segmented in 18 quarters ($Q_0 - Q_{17}$) due to the telescope re-orienting itself every 90 days. The *Kepler* observing quarters resulted in Q_0 of 10 days, and Q_1 of 33 days for the commissioning phase and, segmented 90-day LCs for Q_2 to Q_{16} (see public data release 25, Thompson et al. 2016; Van Cleve and Caldwell 2016). The second month of Q_{17} was terminated after less than five days of observation, following the failure of reaction wheel 4.

Sample A (see Figure 4.1 panel II) was selected to test the performance of the GPS method for stars with near-solar rotation periods. Hence, in this sample, we considered stars with a narrow range of rotational periods between 24.0 and 27.4 days (i.e. encompassing the sidereal Carrington rotation period of the Sun at 25.4 days) and a broad range of effective temperatures $T_{\text{eff}} \in (5000\text{--}6000)$ K. These selection criteria yielded a sample consisting of 686 stars. Among this sample, 282 stars also have rotation periods obtained by Reinhold and Gizon (2015) using the generalised Lomb-Scargle periodograms (hereafter, GLS).

Sample B (see Figure 4.1, panel III) was selected to study the dependence of the inflection point position on the rotation period. Therefore, in contrast to Sample A, we considered stars with a broad range of rotation periods (between 15 to 40 days) but a narrow range of effective temperatures (5700–5900 K, encompassing the solar value of 5778 K). These criteria led to the selection of 361 stars for Sample B after removing the overlapping targets the with initial Sample A. Rotation periods of 172 stars in this sample were also reported in Reinhold and Gizon (2015).

Between our two samples, we thus considered 1047 *Kepler* stars in all. The LCs were acquired in the long-cadence mode (i.e. with a cadence of 29.42 min). Following McQuillan et al. (2014) and Reinhold and Gizon (2015), we utilised LCs from $Q_1 - Q_{14}$ processed with the pre-search conditioning and Bayesian maximum a posteriori approach (PDC-MAP, see Smith et al. 2012). For quarters $Q_{15} - Q_{17}$, only processing with multiscale MAP (PDC-msMAP Stumpe et al. 2014) is available.

In Figure 4.2, we plot the distribution of variability ranges in our samples A and B. These variability values are defined by computing the difference between the 95th and 5th percentiles of the sorted flux values for each of the *Kepler* observing quarters (see Basri et al. 2011) and then taking the median value among the quarters. This defined variability range was chosen versus the approaches based on the standard deviation analysis by Mathur et al. (2014); He et al. (2015) or the smoothed amplitude (10th to 90th) method presented in Douglas et al. (2017), given the higher range of amplitude used in Basri et al. (2011). The selection of the methods mentioned are not expected to compromise the analysed outcome. Additionally, we show solar variability ranges computed using total solar irradiance data (TSI, i.e. total radiative flux from the Sun at 1 A.U.) for 1996–2017 obtained by the *Variability of solar IRradiance and Gravity Oscillations* (VIRGO; Fröhlich et al. 1997) experiment on the *SOlar and Heliospheric Observatory* SoHO mission. For these VIRGO data, the entire 1996–2017 observation period was split into 6 Kepler-like time ranges (five 1530-day periods and one 712-day period for a total of 7787-days starting 28 January 1996; see Fig. 3 from Paper II). The solar variability value for each of the time ranges is represented in Fig. 4.2 as vertical green dashed lines. This gives a range of solar variability of $Var_{\odot} \in (400\text{--}1300)$ ppm.

Figure 4.2 shows that most of the stars in our samples are much more variable than the Sun. This agrees with García et al. (2014); Buzasi et al. (2016b); Reinhold et al. (2020b), who showed that solar-type stars (i.e. stars with near-solar fundamental parameters and rotation periods) are, on average, significantly more variable than the Sun. Furthermore, our samples A and B also contain stars which are cooler and rotate faster than the Sun. These stars are also expected to be more variable than the Sun (see e.g. McQuillan et al. 2014, for a discussion of the dependence of the variability on the rotation period and temperature).

We note that an anomalous low variability of solar-type stars found by Reinhold et al. (2020b) does not necessarily imply that the Sun is an outlier. An alternative explanation is

4 Inflection point in the power spectrum of stellar brightness variations III: Facular versus spot dominance on stars with known rotation periods

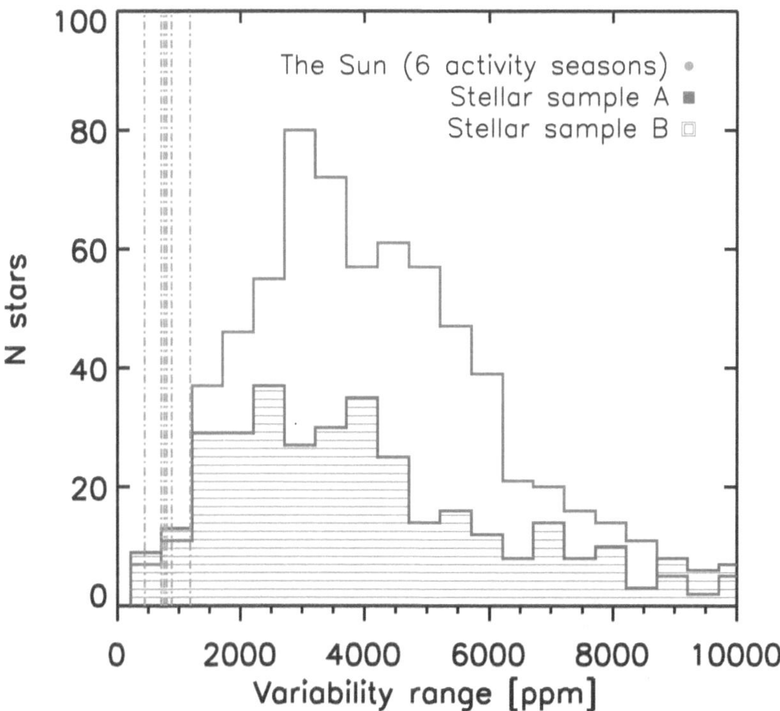

Figure 4.2: Histograms for variability ranges from samples A (violet outline) and B (blue rectangles). Vertical green dashed lines represent solar variability calculated for 6 activity seasons using 21 years of VIRGO TSI data from Paper II (see text for more details).

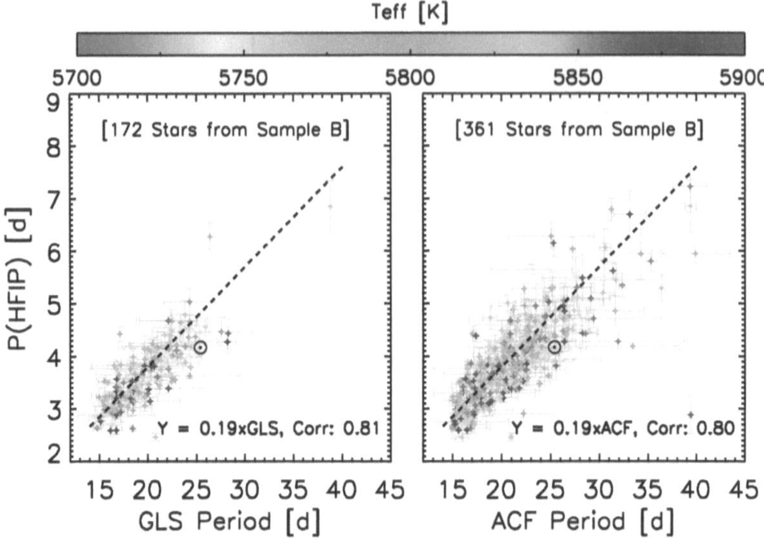

Figure 4.3: Only sample B is shown here. Position of the high-frequency inflection point (P_{HFIP}) is plotted against rotation period. Rotation periods are taken from Reinhold and Gizon (2015) (left panel) and McQuillan et al. (2014) (right panel). Colours represent the stellar effective temperature, T_{eff}. The Sun is represented by the solar symbol $\odot$. Dashed lines in both panels indicate a linear fit constrained to go through the origin of the coordinate system. A logarithmic visualization is available in the appendix, see Fig. 4.6.

that by comparing solar variability to the sample of stars with known rotation periods, we focus only on a small sub-sample of stars for which the ACF method could return rotation periods (and the Sun most probably would not belong to such a sample). Along this line of reasoning, Reinhold et al. (2020b) found that solar levels of photometric variability are typical for stars having near-solar fundamental parameters but unknown rotation periods.

4.3   Results and discussion

In this section, we calculate the position of the inflection point for each star in the samples A and B defined in Sect. 4.2. Following the methodology described in Papers I and II, we first calculate the power spectra of the stellar brightness variations using a Paul wavelet on the order of six (see Torrence and Compo 1998) for the *Kepler* observing quarters $Q_1 - Q_{17}$. We determined the period corresponding to the high-frequency inflection point, $P_{\mathrm{HFIP}_{(Q_n)}}$, per quarter and calculate the mean value for P_{HFIP} over all 17 quarters for each star. This allows us to obtain a unique representative value of P_{HFIP} per star. The uncertainty is calculated using 2-σ of the distribution of the obtained P_{HFIP} values. Finally, we used the P_{HFIP} to calculate the stellar rotation period, P_{rot} (see Table 4.2 and on-line reference for a compilation of GPS outputs and comparison with GLS and ACF reference values).

4 Inflection point in the power spectrum of stellar brightness variations III: Facular versus spot dominance on stars with known rotation periods

In Figure 4.3, we plot the mean values of the P_{HFIP} positions for each of the stars against the rotation periods from Reinhold and Gizon (2015) (left panel, GLS) and McQuillan et al. (2014) (right panel, ACF). The rotation periods and positions of the inflection points are well-correlated. A linear fit constrained to go through the origin of the coordinate system gives $P_{\text{HFIP}} = 0.19 \times P_{\text{rot}}$ with Pearson coefficients for periods of 0.81 from Reinhold and Gizon (2015) and 0.80 from McQuillan et al. (2014).

The scatter around the linear fits has multiple sources. First, the calibration coefficient between rotation period and inflection point, $\alpha = P_{\text{HFIP}}/P_{\text{rot}}$, depends on the relative roles that bright faculae and dark spots play in generating stellar brightness variations. According to the model presented in Paper I, these roles are regulated by the ratio between facular and spot areas of active regions at the time of emergence, $S_{\text{fac}}/S_{\text{spot}}$ (i.e. zero ratio would lead to a purely spot-dominated star, while very large ratios would correspond to a faculae-dominated star). Secondly, there is an intrinsic statistical uncertainty of the GPS method. For example, in Paper I, we found that even for a star with a fixed $S_{\text{fac}}/S_{\text{spot}}$ ratio, the factor α showed 5-10% variations from one realisation of active regions emergence to another. Finally, there is also an uncertainly in the determination of rotation periods by Reinhold and Gizon (2015) and McQuillan et al. (2014) (see e.g. Fig. 4.7, where we compare the periods from these two sources for the 172 stars of sample B that are common to both).

Table 4.2: GPS outcome values.

KIC	P_{HFIP} [d]	σP_{HFIP} [d]	α	$\sigma \alpha$	P_{rot} GPS [d]	P_{rot} GLS [d]	P_{rot} ACF [d]	log g	[Fe/H]	Var [ppm]	T_{eff} [K]
	[—————————— (1) ——————————]					[– (2) –]	[— (3) —]	[—— (4) ——]			
10070928	3.78	0.108	0.173	0.0050	19.894	22.132	21.747	4.594	-0.46	4688	5706
10080186	3.67	0.131	0.207	0.0074	19.315	18.239	17.747	4.547	-0.10	7472	5749
10080239	2.96	0.084	0.184	0.0052	15.578	16.577	16.131	4.547	-0.14	7174	5792
10083970	3.08	0.097	0.188	0.0059	16.210	16.308	16.374	4.559	-0.20	11418	5745
10089777	3.65	0.066	0.188	0.0034	19.210	19.437	19.381	4.541	0.07	3752	5713
10091612	2.90	0.078	0.130	0.0035	15.263	–	22.214	4.550	-0.18	1760	5804
10125510	3.78	0.180	0.161	0.0076	19.894	–	23.427	4.372	-0.64	0608	5838
10129857	4.69	0.235	0.162	0.0081	24.684	–	28.859	4.536	-0.06	1488	5757
10136417	4.46	0.276	0.169	0.0105	23.473	27.541	26.288	4.303	0.16	2690	5849
10140949	4.12	0.127	0.182	0.0056	21.684	–	22.636	4.501	0.02	1999	5874
10146308	3.69	0.191	0.174	0.0090	19.421	–	21.193	4.591	-0.52	3340	5804
10064358	3.82	0.078	0.231	0.0047	20.105	16.526	16.533	4.486	-0.22	2526	5772

Notes. This table contains an example of the GPS outputs, the compared rotation period values from GLS & ACF, and stellar parameters for 12 randomly selected objects from samples A & B. 1) GPS outcome: In column 2 P_{HFIP} is given, in column 3 its 2-sigma uncertainty, σP_{HFIP}, defined from individual inflection points for each *Kepler* observing quarter. In column 4 and 5 values of α-factor and its 2-sigma uncertainty are reported respectively. P_{rot} GPS values in column 6, as result of applying Eq. 4.1 using the factor $\alpha = 0.19$. 2) Column 7 shows the P_{rot} reported by Reinhold and Gizon (2015). 3) P_{rot} and variability values (Var in [ppm]) reported by McQuillan et al. (2014) in column 8 and 11. 4) Cols. 9, 10 and 12 show the log g, [Fe/H] and T_{eff} respectively, taken from Huber et al. (2014). A complete table for the 1047 objects is available in a machine-readable form in the online journal and at the Centre de Données astronomiques de Strasbourg

4   Inflection point in the power spectrum of stellar brightness variations III: Facular versus spot dominance on stars with known rotation periods

In Fig. 4.4, we show calibration factors, α, for samples A (top panel) and B (bottom panel). The rotation periods of stars in both samples are taken from McQuillan et al. (2014). In Paper I, we demonstrated that the profile of the high-frequency tail of the power spectrum and, consequently, the values of α depend on the center-to-limb variations (CLVs) of the brightness contrasts of magnetic features. Since spots and faculae have different CLVs, the value of α depends on their relative contributions to the stellar brightness variations. For the extreme cases, we found that α is about 0.14 for simulated stellar light curves with variability solely determined by faculae and about 0.21 for simulated stars with variability dominated by spots. These values are designated, respectively, by the red and green horizontal dashed lines in Fig. 4.4. It is reassuring to see that most of the α values for samples A and B appear between these two extreme-cases. Stars with values of α outside of this range (in particular, with $\alpha > 0.21$) are likely due to: inclination angles below $45°$, which can lead to a shift of the inflection point to lower frequencies (see Fig. 9 from Paper I); statistical noise of the GPS method; and possible uncertainties in rotation periods from McQuillan et al. (2014).

For sample A, the ratios are shown as a function of stellar effective temperature from Huber et al. (2014), while for sample B, they are plotted as a function of stellar rotation period from McQuillan et al. (2014). The upper panel of Fig. 4.4 shows that for near-solar rotation periods (the rotation periods in sample A were constrained between 24 and 27.4 days; see Table 4.1), the position of the inflection point shows no significant dependence on the effective temperature (e.g. the fitting of a slope gives a value of 7.36×10^{-7}, which is well below the $1\,\sigma$ uncertainty of 1.8×10^{-6}). We also note that the mean value of $\alpha = 0.19$ is equal to the slope of the regression shown in Fig. 4.3. This implies that neither the $S_{\mathrm{fac}}/S_{\mathrm{spot}}$ value nor CLVs of facular and spot contrast change significantly within the 5000-6000 K domain of sample A. We note, however, that we cannot conclusively exclude the improbable scenario that the effect from the change of the facular and spot contributions to brightness variability on α is compensated by a change of facular and spot CLVs, such that the net effect on the inflection point is very small.

The bottom panel of Fig. 4.4 shows that for stars with near-solar effective temperatures there is a rather weak but statistically significant dependence of the α factor on the rotation period. For example, fitting a linear dependence returns a slope value of 9.3×10^{-4} which is 3.8 times larger than its 1σ uncertainty of 2.5×10^{-4}. However, the value of the slope is strongly affected by a couple of slowly rotating stars and, thus, might not represent a trend in the full sample. To better characterise such a trend, we calculated the mean value of the calibration factor in several bins of rotation period values. We compiled the mean α values per several bins of rotation periods; see Table 4.3 for details. To further illustrate the trend of α values with rotation period, the histogram to the right side of the panel shows the distributions of α values for two rotation periods - one for stars with rotation periods below 21 days and another with rotation periods above 21 days. We can see that the two distributions are clearly shifted relative to each other and the α-values of faster rotating stars are larger than those slow-rotating stars.

We note that the 'n' number of stars and the amplitude of photometric variability in our samples decreases with rotation period. Consequently, slow rotators might be more affected by photometric noise. We investigated the possible effect of *Kepler* white noise on

---

[1]   A logarithmic visualization of the relation α factor versus P_{rot} is available in the appendix, see Fig. 4.8.

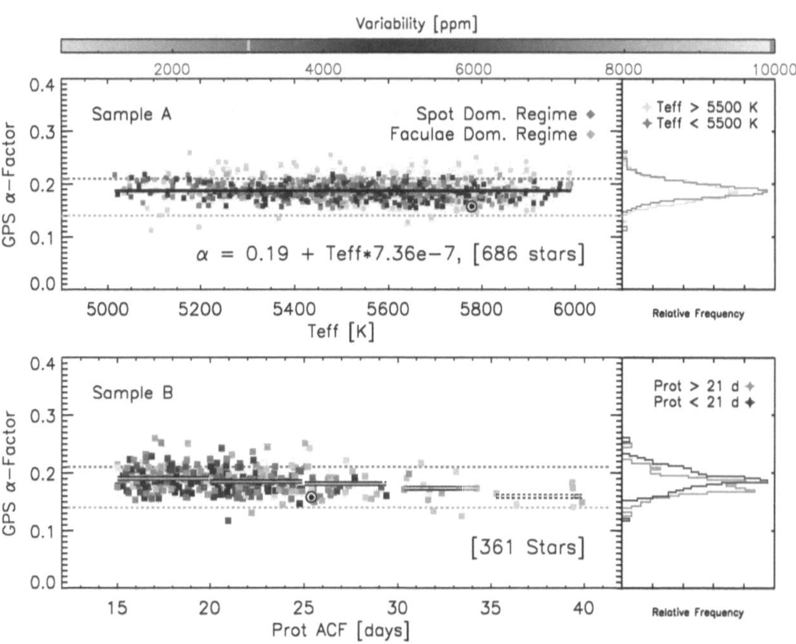

Figure 4.4: Top panel: α factor versus T_{eff} for Sample A shows consistency across a broad temperature range. The black line corresponds to the linear fit to values with an uncertainty within 2-σ of the mean of the distribution as shown by coloured squares (grey squares lie outside the 2-σ of the distribution). The histograms to the right side of the panel display the distribution of α values for two effective temperature regimes, using $T_{\text{eff}} = 5500$ K as a threshold. Bottom panel: α factor versus P_{rot} from McQuillan et al. (2014) for Sample B shows a slight decrease in α with rotation period. The coloured segments indicate the mean of α for the different P_{rot} ranges as indicated in Table 4.3. The histograms to the right of the panel indicate the distribution of α values for two rotation period regimens, using 21 d as a threshold. For both panels the error bars represent 2-σ uncertainties of the α values over all *Kepler* quarters available for each star. The gray squares lie outside of a 2-σ of the distribution. The dashed red and green horizontal lines represent the α factor values in the extreme cases with all variability being due to spots ($\alpha = 0.21$) and all due to faculae ($\alpha = 0.14$), respectively [1].

the deduced positions of inflection points for the stars in our samples. In Fig. 4.5, we plot the dependence of the α factor values on the expected *Kepler* noise levels for each of the stars, calculating the amplitude of the *Kepler* noise as a function of the *Kepler* magnitude (following Lammer 2013). The derived precision (called the noise in the context of *Kepler*) accounts for noise introduced by the instrument and gives it as a function of the *Kepler* magnitude of the source and the variability of sources (see Fig 4.9).

As much as 99.9% of the stars in our sample present a *Kepler* magnitude of 16 mag

4 Inflection point in the power spectrum of stellar brightness variations III: Facular versus spot dominance on stars with known rotation periods

Table 4.3: Mean α-values in sample B per bin.

Bin	n	P_{HFIP} [d]	$\bar{\alpha}$	σ	$\sigma/\sqrt{n}$	Bin colour
1	158	[14–20]	0.190	0.001	9.0×10^{-5}	Pink
2	148	[20–25]	0.184	0.001	9.8×10^{-5}	Purple
3	36	[25–30]	0.181	0.003	5.0×10^{-4}	Orange
4	12	[30–35]	0.173	0.006	1.9×10^{-3}	Gold
5	6	[35–40]	0.156	0.004	0.17×10^{-3}	Yellow

Notes. Compilation of mean α-values for n stars per range of rotation periods, see Fig. 4.4.

or fainter. We find that values of the α factor are independent of the *Kepler* noise, with fits of a linear dependence to samples A and B giving slope values well below their 1σ uncertainties (7.4×10^{-7} and 4.9×10^{-8}, respectively). Consequently, we do not expect the *Kepler* noise to affect the positions of inflection points determined for stars in our samples. Furthermore, we note that photometric noise would shift the position of the inflection point to lower frequencies (see Paper II), meaning that it would lead to a trend that is opposite to what we see in the bottom panel of Fig. 4.4.

A possible explanation of the observed tendency in Fig. 4.4 is a change in the relative contribution of faculae and spots to stellar rotation variability (or S_{fac}/S_{spot} ratio in terms of Paper I) with rotation period. The increase of the α factor with rotation rate implies that the S_{fac}/S_{spot} ratio (and, consequently, the contribution of faculae to the rotational variability) is lower in faster rotating and, therefore, more active stars. Such a trend is consistent with an extrapolation to higher activities of observed solar behaviour. Indeed, the mean size of spots on the Sun increases during periods of high solar activity (Hathaway 2015; Mandal et al. 2020). At the same time the S_{fac}/S_{spot} ratio decreases with the size of active regions and their spot components. An extrapolation of these trends to activity levels higher than seen in the Sun results in an increase of the α factor with activity, and, consequently, with rotation rate, as indicated by the bottom panel of Fig. 4.4.

We note that the ratio S_{fac}/S_{spot} between facular and spot areas of the individual magnetic features at the moment of their emergence discussed until now is different from the ratio between 'instantaneous' stellar disk coverage by faculae and spots. The former is a property of a magnetic feature during its emergence onto the surface of the star, while the latter is strongly affected by the evolution of the magnetic flux after emergence. For example, in the hypothetical case of facular portions of active regions evolving exactly as spot portions, these two ratio remain the same. In reality, the 'instantaneous' ratio is generally significantly larger than that 'at the time of emergence' since faculae live longer than spots.

Solar observations show that the ratio between such instantaneous solar-disk coverage by faculae and spots decreases as solar activity increases (Chapman et al. 1997; Foukal 1998). The observed patterns of stellar-brightness variability indicate that this trend also extends to activity values that are significantly higher than those observed on the Sun (Shapiro et al. 2014b). Our result indicates that not only the ratio between instantaneous facular and spot disc coverage shows this trend. Also facular to spot area ratio corresponding to individual active regions 'at the time of emergence' continues to decrease with increasing level of activity, also beyond the level of solar activity observed until now. We note that this result is not a simple consequence of the drop in the instantaneous ratio. Simulations with a surface flux transport model by Cameron et al. (2010) show that the origin of the decrease in the instantaneous ratio with increasing activity is rather complex. It is, to a great extent, caused by a stronger cancellation of small-scale magnetic field associated with faculae. Consequently, it does not necessarily demand any changes in the structure of the emerging magnetic flux which defines the ratio corresponding to the individual active region at the time of emergence; see the discussion in paper I.

The bottom panel of Fig. 4.4 shows that the dependence of α on the rotation period is quite noisy, that is, there is quite a large spread of values for a fixed rotation period. This spread basically covers the entire range of values between faculae- and spot-dominated variability. In particular, it is significantly larger than statistical noise in the inflection point position that we found in Paper I. We speculate that such a large spread implies that the $S_{\mathrm{fac}}/S_{\mathrm{spot}}$ ratio is not uniquely defined by the stellar effective temperature and rotation period.

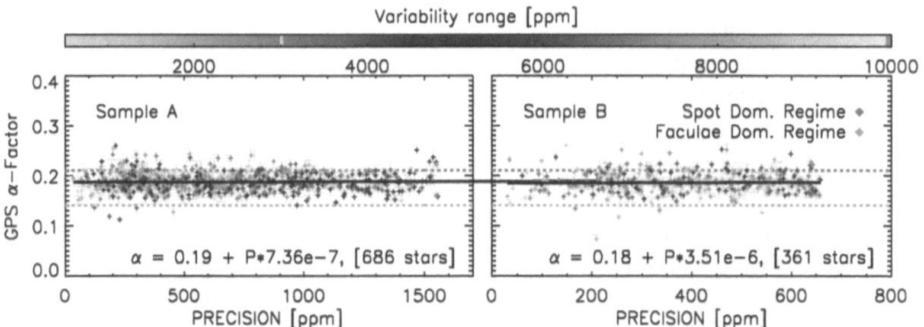

Figure 4.5: α factor versus photometric precision for sample A (left panel) and sample B (right panel) for *Kepler* observations. Those records for single long observation of isolated stars observed in an uncrowded pixel, i.e light curves of resolved targets without contamination of additional sources. Extreme-case limits for spot- and faculae-dominated stars are shown as horizontal dashed lines in red and green, respectively. The individual data points are coloured according to the detected variability range for that particular star. This is consistent with previous figure 4.4. The error bars represent 2-σ uncertainties of the α values distribution over all *Kepler* quarters available per star. Gray rhomboids represent data points that lie more than 2-σ from the centre of the distribution.

In Papers I and II, we found that the solar value of the calibration factor ($\alpha_{\mathrm{Sun}} = 0.158$) is closer to the faculae-dominated case ($\alpha = 0.14$) than to the spot-dominated case ($\alpha = 0.21$). Interestingly, Fig. 4.4 shows that the solar (α_{Sun}) value appears to be rather low relative to that of stars in both of our samples (see also Fig. 4.3, where the Sun is clearly below the regression line). This is, however, not surprising since most of the stars in our samples are significantly more variable than the Sun even though we selected the stellar sample by the reported detected rotation period and not by the variability (see Fig. 4.2, cf. Reinhold et al. 2020b). This implies that these stars are also more active than the Sun (see also Zhang et al. 2020, who showed that stars with known near-solar rotation periods have systematically higher values of S-index than the Sun). Therefore, we can expect that their $S_\mathrm{fac}/S_\mathrm{spot}$ ratios are smaller and, consequently, their α factors are larger.

4.4   Summary

In this study, we developed the GPS method, which is a novel means for determining stellar rotation periods from photometric time series. Instead of basing this determination on the more traditional means of identifying the strongest peak in a Lomb-Scargle periodogram or a maximum of the auto-correlation function, we identify the steepest point (i.e. the inflection point) in the global wavelet power spectrum of stellar brightness variations. In Paper II, we showed that while the solar brightness contributions from faculae and spots can oppose each other to reduce any peak due to the rotation period from Lomb-Scargle periodograms and auto-correlation functions, it has only a very minor effect on the location of this high-frequency inflection point and the resulting ratio between the period corresponding to the inflection point and the actual rotation period, α. In Paper I, the factor of α, however, shows a moderate dependence on the relative contribution of faculae to stellar brightness variations. Therefore, identifying the position of the inflection point allowed determination of the rotation period in stars where other methods fail (with an internal uncertainty of about 25%). At the same time, this GPS method allows us to assess the relative role of faculae in stars with known rotation periods. In Paper II, we tested the performance of the GPS method against solar photometric data. We demonstrated that in contrast to other methods, the GPS method allows for an accurate determination of the solar rotation period independently of the solar activity level.

In this study, we applied the GPS method to 1047 F-, G-, and K-type stars with rotation periods as reported in McQuillan et al. (2014). We show that the position of the high-frequency inflection point is well-correlated with the rotation periods of stars in the two samples we analysed, providing further validation of the GPS method. We emphasise that the stellar light curves analysed in this study and the solar light curves analysed in Paper II are quite different: the amplitudes of brightness variability in the stellar samples in this study are generally higher than that of the Sun, and the stellar brightness modulation are much more regular over rotational timescales.

We find that the α factor increases with rotation rate, indicating that faculae become less important on stars rotating faster than the Sun. We have also found that the facular contribution to solar brightness variability is larger than its contribution to brightness variability in a sample of stars having near-solar rotation periods and temperatures. We attribute this to a selection effect since the rotation periods of stars with brightness-variability patterns similar to that of the Sun are rather difficult to measure via the ACF method for rotation-period determinations and, thus, there is a dearth of such stars in our sample. Consequently, our results indicate that in addition to being more active than the Sun (see also Reinhold et al. 2020b; Zhang et al. 2020), the stars with near-solar effective temperatures and near-solar rotation periods determined by McQuillan et al. (2014) have different compositions of active regions (with smaller facular contributions).

The GPS method for determining rotation periods could thus prove to be an important contributor to enhance the lower-than-expected number of G-type stars with near-solar rotation periods reported by van Saders et al. (2019). This method can also improve the solar-stellar comparison as in Reinhold et al. (2020b). The outcome of GPS might bring a new perspective in understanding stellar activity.

While we focus in this study on applying GPS to stars with known rotation periods, in a forthcoming study, we plan to apply the GPS method to *Kepler* stars with previously unknown rotation periods as well as to *TESS* stars. This might help establish a new and more complete sample of stars having near-solar rotation periods, based upon which we can investigate whether solar variability still appears anomalously low in comparison to stars in this broader sample.

This will be of importance to the exoplanet community, since the knowledge of rotation periods will help identify radial velocity jitter from planetary signals (see Oshagh 2018; Faria et al. 2020; Hojjatpanah et al. 2020). The anticipated GPS-determined expanded database of stellar rotation periods could also bring crucial information for ongoing and upcoming surveys such as NIRPS-HARPS and ESPRESSO (see Pepe et al. 2010; Bouchy and Doyon 2018). Additionally, the precise determination of host-star rotation periods is important for recovering accurate exoplanet radii, which will be crucial for future searches of transiting Earths or super-Earths in the light curves of solar twins in the PLATO field (see Rauer et al. 2014).

4.5   Appendix chapter 4

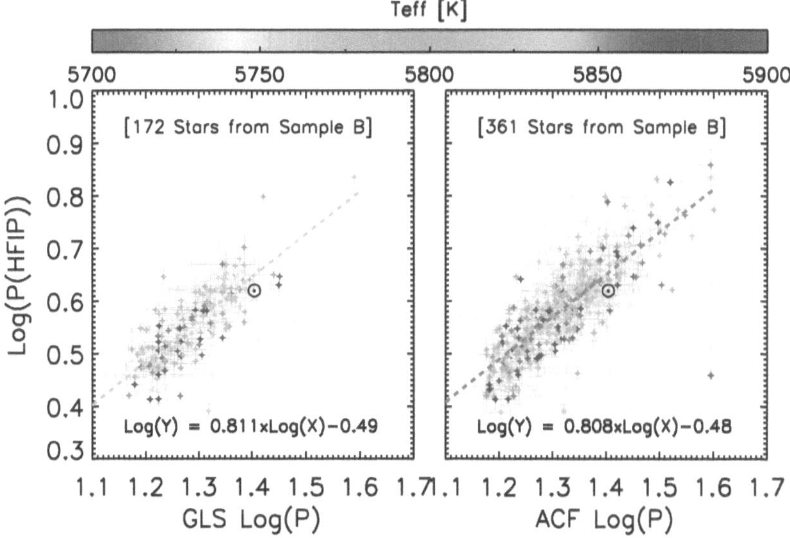

Figure 4.6: Logarithmic visualization of Fig. 4.3.

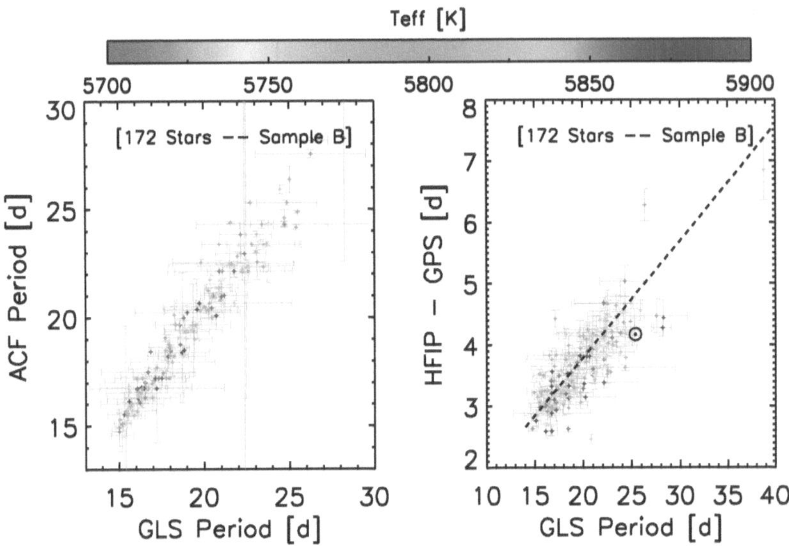

Figure 4.7: Comparison for the 172 stars in sample B with reported rotation periods by the ACF, GLS and GPS methods. Left: ACF versus GLS. Right: HFIP-GPS versus GLS. Scatter colour to visualise the temperature. The comparison is made in a similar range scale for a better visualization.

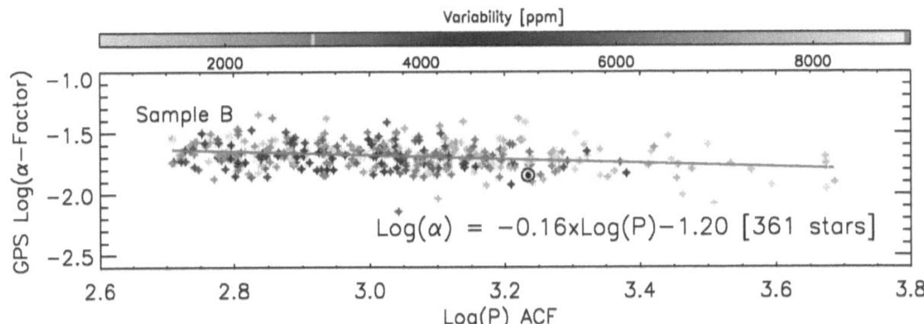

Figure 4.8: Logarithmic visualization of bottom panel in Figure 4.4.

4 Inflection point in the power spectrum of stellar brightness variations III: Facular versus spot dominance on stars with known rotation periods

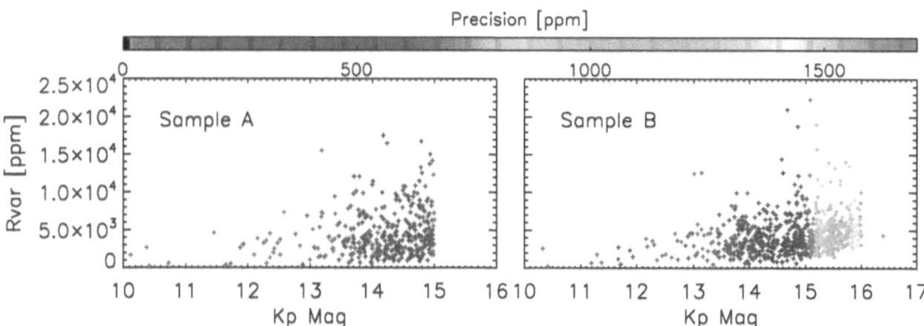

Figure 4.9: Variability range in ppm versus Kepler magnitude (Kmag). The colour bar indicates Kepler precision.

Summary & outlook

The complexity of solar-like brightness variability has been an issue for attaining confident rotation period values, one of the most important parameters for stellar analysis. The low amplitude signal and the irregular modulation in the light-curves are particular from stars as the Sun. Those characteristics make difficult to fit a sinusoidal function to the signal and, are the main reason of why most of the common methods used to recover rotational velocities fail. In this work the GPS method, a novel tool for recovering rotation periods, was successfully proposed, developed, tested, and implemented. The GPS method performed well on simulated and observed solar and stellar light-curves.

In Chapter 1.1 an introduction about the stellar rotation phenomena is contextualized to the topics treated in this book. A basic drawn of the stellar rotation evolution and related topics such as the rotation-age connection, the Skumanich law, and its deviations are mentioned. As well, is pictured the connection between rotational velocity and activity. Basic information about the solar and stellar photometric time-series, the instruments, and the methods utilized to perform this work are described. The the introduction is complemented with a brief compilation of the state of the art.

Sections (1.7, 1.8 and, 1.9) contain the compilation of three additional publications were I contributed as co-author and, the GPS method was successfully implemented. The abstracts and main contributions in the three additional co-authored publications were included in the Introduction as supplementary material. This, given their pertinent correlation with the present book. In the section 1.7, is performed a in-deep analysis and comparison between solar and stellar variability. This analysis concluded that the Sun show lower activity signatures on its light-curves in comparison with their stellar analogous. The results of this paper is coherent with the analysis and discussion performed in this book. In the section 1.8 the GPS is successfully applied on a *TESS* light-curve for the star HD 41248. In the section 1.9 GPS is applied for a sample of 171 *TESS* stars. Rotation periods and facular to spot ratio are reported for 71 and 30 stars of the sample, respectively.

The core of the work presented in this book has been segmented in three main chapters (2, 3 and, 4) based on 2 published and 1 accepted manuscripts at A&A journal. Chapter 2 is shown an extensive analysis of the gradient of the power spectra of Solar-like simulated brightness variations time-series. Different models of synthetic data are used to describe the possible magnetic feature configurations that compose stellar light-curves. The gradient of the power spectra is applied to those data, analysing the different peaks obtained (inflection points). There are two main inflection points that are recurrent and show a strong amplitude in most of the LCs realizations. Those inflection points present a consistent proportionality with the input rotation velocity and the magnetic feature configuration. GPS, the novel method for retrieving rotation periods on stars alike the Sun is based on the characterization of the proportionalities found.

After analyse the relation between the inflection points located with GPS and the input values of rotation period on the synthetic LCs, a validation with Solar brightness variations was performed in order to corroborate such proportionality on observational data, see Chapter 3. The observations analyzed included two independent data-sets recorded by VIRGO/SoHO and TIM/SORCE. The selection of two different sources allowed to discard possible superposed instrumental systematics over the period of the inflection points. Using the different TSI time-series with simultaneous MDI/SoHO images was possible to pseudo-isolate and characterize the related effect of spots and faculae over the light-curves. By using this information and comparison with the synthetic light-curves in Chapter2, was possible to recognize particular signatures of the dominance of faculae or spots presence using the GPS.

In a subsequently step, after having a positive outcome testing the GPS with solar data, in Chapter 4 was performed a co-validation of the method proposed in this book using *Kepler* observations. There is a comparison of rotation period outputs by ACF, GLS, PS and GPS for about 1000 stars. The rotation periods of those stars ware reported previously on different catalogs, and then used in this work as benchmark for testing the GPS. Even more than just validate the GPS against other methods, in Chapter 4 the relation between spot or faculae dominance and the location of the main inflection point from the GPS is confirmed for active stars in agreement with Chapter 2.

Applying the GPS over a extended sample of star in the *Kepler, TESS & PLATO* fields is suggested as additional and complementary work.

The contributions of this book to stellar Astrophysics encompasses the possibility of analyse rotation period on star like the Sun and, for the first time, have an estimation of the ratio between facular to spots components expressed in the time-series of the brightness variations. The stellar characterization by GPS have the potential to expand our knowledge about the closest solar analogues.

www.ingramcontent.com/pod-product-compliance
Lightning Source LLC
LaVergne TN
LVHW041710190726
843493LV00007B/2024